Scheduled Technical Proceedings

of the

Estes Park
Advanced Propulsion Workshop

19 – 22 September 2016

Estes Park, Colorado, USA

edited by

H. Fearn   L. L. Williams

CSU Fullerton - Physics   *Konfluence Research Institute*

Scheduled Technical Proceedings
of the Estes Park Advanced Propulsion Workshop

edited by H. Fearn & L.L. Williams

This content in print or digital media is made available in 2016 under the terms of the Creative Commons Attribution & ShareAlike 4.0 License (CC BY SA), https://creativecommons.org/licenses/by-sa/4.0/. Further distribution of this work must maintain attribution to the author(s), the title, and the name of this Proceedings.

ISBN: 978-0-9753995-5-2
Library of Congress Control Number: 2016961880
Konfluence Press
Manitou Springs, Colorado

Dedicated to

JAMES F. WOODWARD

on the occasion of his seventy-fifth birthday

ESTES PARK ADVANCED PROPULSION WORKSHOP

Held at the YMCA of the Rockies
Estes Park, Colorado USA
19 –22 September 2016

Organizing and Technical Committee:

H. FEARN (Cal. State Univ. Fullerton – Physics)
L.L. WILLIAMS (Konfluence Research Institute)
J.F. WOODWARD (Cal. State Univ. Fullerton – Physics)

Workshop Production:

R. SMITH (audio)
R. SNELSON (video)
H. FEARN (moderator)
L.L. WILLIAMS (moderator)
&
the staff of
The YMCA of the Rockies

Workshop Sponsorship:

The Space Studies Institute

TABLE OF CONTENTS

PUBLISHER'S NOTE 3

PARTICIPANTS 4

GROUP PHOTO 5

SCHEDULED TECHNICAL AGENDA 6

ETHOS OF WORKSHOP 8

GENERAL SESSION:

    The Fuel Problem (Quick Study I) 10
    The Time-Distance Problem (Quick Study II) 11
    Warp Drives & Wormholes: Good News, Bad News (QS III) 13

    General Session Prep Summary 15

    Aspects of Plausible Extensions to Physical Law – *L.L. Williams* 17
    Experimenting with Novel Propulsion Ideas – *G. Hathaway* 23

MACH EFFECT THRUSTER SESSIONS:

    Maxwellian Gravity (QS IV) 49
    Inertia from Gravity: Insights of Sciama (QS V) 51
    Woodward's Mach Effect Equation (QS VI) 54

    Mach Session Prep Summary 57

    Gravitational Absorber Theory & The Mach Effect – *H. Fearn* .. 59
    Thrust Signature of a Mach Effect Device – *N. Buldrini* 91
    A Conventional Post-Newtonian Mach Effect – *L.L. Williams* 99
    Mach Effect Discussion with J.F. Woodward 102
    Mach Effect Discussion with J.A. Rodal 104
    Mach Effect Group Discussion 107

RF Cavity Thruster Sessions:

RF Cavity Thruster Prep Summary 122
Experiments with RF Cavity Thrusters – *P. March* 123
Revolutionary Propulsion Research at TU Dresden – *M. Tajmar* 163
GEM Theory of Q-V Thruster – *J. Brandenburg* 187
EM-drive Based on Mach-Lorentz Theory – *J.-P. Montillet* 209

Kaluza Session:

Kaluza Session Prep Summary 226
Verification of the Kaluza Theory – *L.L. Williams* 227

Chameleon Cosmology Session:

Chameleon Session Prep Summary 237
Applications of Chameleon Cosmology – *G.A. Robertson* 239

Tri-Space Session:

Tri–Space Session Prep Summary 260
Tri–Space Model of Spacetime and the Universe – *G. Meholic* .. 261

Observations & Summing Up – *L.L. Williams & H. Fearn* 267

Pre-Conference Overview and Mission Statement 271

Publisher's Note

This volume contains the Scheduled Technical Proceedings of the 2016 Estes Park Advanced Propulsion Workshop. Our mission statement, sent out to participants during the workshop planning, is included in the back of this volume. We hoped to bring a new level of scientific and engineering rigor to a field where enthusiasm is sometimes mistaken for engineering, and where breakthrough propulsion conferences are crowned with sci-fi author panels.

Toward that end, the workshop was modeled on the famous Shelter Island Conference, with a small group of invitees in a resort setting. A set of session topics and session leaders were assigned and scheduled by the workshop technical committee. The meeting was intentionally scaled to keep all sessions plenary. Preparatory materials were provided by session leaders and sent to all participants in advance, for each session. Whiteboards were available during sessions, and session leaders accommodated detailed technical interchange with the audience.

The workshop was also organized in celebration of the 75th birthday of Professor James Woodward. There were many celebration activities and unscheduled presentations, in addition to the scheduled agenda. The Space Studies Institute has released a commemorative proceedings of the workshop, in hardcover and eBook. It includes all scheduled and unscheduled presentations, an attached monograph, and an account of workshop festivities. This volume is an abridged version that gathers together only the scheduled technical proceedings.

This focus volume is offered to capture the assessments in detail, and to offer an example of scrutinizing a putative breakthrough in propulsion by preparing a small group of people for in-depth technical review sessions. A wide net was thrown in the selection of sessions; we hoped to distinguish our workshop in how we treated them. The reader finds here a sequence of prep material, proceedings papers, and recorded discussion, for the scheduled technical sessions. It is intended to be sufficient for the reader to determine whether, and how, further investigation may be warranted for the concepts considered.

L.L. Williams
Manitou Springs, Colorado
February 2017

Participants

John Brandenburg
Michelle Broyles
Nembo Buldrini
Dennis Bushnell
Bill Christie
John Cole
Todd Desiato
Heidi Fearn
Jeremiah Hansen
Jan Harzen
George Hathaway
Gary Hudson
Eric Jansson
Peter Jansson
David Jenkins
Wes Kelly
Eric Laursen
Anthony Longman
Paul March
David Mathes
Greg Meholic
Jean–Philippe Montillet (via Skype)
Glen Robertson
José Rodal
Martin Tajmar
Ron Turner
Lance Williams
Jim Woodward

Group Photo: Participants of the Estes Park Advanced Propulsion Workshop. Seated at the front, left to right James F. Woodward and Heidi Fearn. From the left standing, Lance Williams, Jeremiah Hansen, Jan Harzen, Nembo Buldrini. Front row, Greg Meholic, John Cole, Tony Robertson, Martin Tajmar, Anne Hudson, Gary Hudson, José Rodal and Michelle Broyles. Back row from the left, Peter Jansson, David Jenkins, Ron Turner, Eric Laursen, David Mathes, Anthony Longman, Eric Jansson (Obscured), Bill Christie. On the right behind Michelle is Dennis Bushnell, Paul March, Todd Desiato, and Wes Kelly.

Estes Park Advanced Propulsion Workshop
19 – 22 September 2016
Scheduled Technical Agenda (16 September)

Topic area in bold, *specific concept in italics*, presenter named.
Four blocks per day up to 100 min each.

Evening 0: 19 September

- *GEM theory mixer* – John

Day 1: 20 September

- Block 1: **General**
  - *Introduction to SSI* – Gary
  - *Ethos of Meeting* – Lance, Jim, & Heidi
  - *Facets of a Valid Extension to the Laws of Physics* – Lance

- Block 2: **General**
  - *Facets of Valid Experiments* – George

- Block 3: **Machian Approaches**
  - *Hoyle–Narlikar Theory & the Mach Effect* – Heidi & José
  - *EM Drive Explained with Mach Effect Theory* – Jean-Philippe

- Block 4: **Scalar Field Approaches**
  - *Kaluza Unification of Gravity and EM* – Lance

Day 2: 21 September

- Block 1: **Electromagnetic Approaches**
  - *Experiments with RF Resonant Cavity Thrusters* – Paul

- Block 2: **Electromagnetic Approaches**
  - *Experiments at Dresden* – Martin

- Blocks 3 & 4: **Free Time**

Day 3: 22 September

- Block 1: **Scalar Field Approaches**
    - *Experimental Application of Chameleon Cosmology* – Tony
- Block 2: **Alternative Conceptions**
    - *Tri–Space View of the Universe* – Greg
- Block 3: **Machian Approaches**
    - *Mach Effects: Concepts and Experiment* – Jim
    - *Mach Effect Experiments from Austria* – Nembo
- Block 4: **Workshop Wrap**
    - *Proceedings* – Heidi & Lance

Ethos of the Workshop

L. L. Williams, H. Fearn, & J. F. Woodward

– WHAT WE HOPE TO DO DIFFERENTLY HERE, AND WHY –

The technical committee welcomes everyone to the workshop. At this workshop, we hope to do something a little different than other science meetings. Please allow us to challenge you – and challenge us! At typical meetings, someone presents a 20 minute paper, with 5 minutes for questions. Then they go home, and repeat next year.

The scope of our ambition demands more time for technical interchange. The format of typical science meetings just does not allow the necessary in-depth scrutiny of an idea among technical peers. This scrutiny is necessary for the community to identify plausible candidate ideas, and also for someone with an idea to explain it to peers and win converts.

As we contemplate this workshop, we take as axiomatic that no one gets to the stars alone. The dream of interstellar travel, if it is to be realized, will entail people working together. We can imagine an industrial enterprise of some sort, built on an engineering discipline. It will entail a broader mainstream science than we know today – a breakthrough will have occurred.

The first step along this road is for the discoverer to convince the second person. If someone has a potential discovery, the first step to bringing in society and mainstream science, is to convince the second person.

If someone has made a discovery, but has not convinced anyone or shared it with anyone, then no contribution is made to society. It might be due to poor people skills – or perhaps someone was delusional all along. The test of sanity is on the anvil of peer review.

Therefore, this workshop is your chance to convince the second person. This is your chance to convince educated peers and move the ball forward for society. And we are providing sufficient time to do it.

As we pursue this common objective, we also take it as axiomatic that we must abide the norms of science, something history shows us is necessary for technical progress of any sort. Some of the propulsion conferences these days degenerate into performance artists and sci-fi author panels. We want to inject a dose of scientific method back into this business.

We want to treat this endeavor with the same rigor and detachment that we would in trying to find a cure for polio, for example: no one looks to UFOs for a cure for polio; no one looks to YouTube for a cure; no one looks to cable news talking heads; no one convenes medical fiction authors to ask what they think. On the contrary, our road to a breakthrough in propulsion proceeds through the ground of repeatable experiment and peer review. Come as a scientist and as an engineer to Estes Park!

We must accord with the pillars of science so far. Concepts like conservation of energy are not easily forfeited. For example, in the 1930s, beta decay seemed to indicated a violation of conservation of energy. Pauli suggested an unseen particle – the neutrino – carried the energy away. If you are invoking the tooth fairy or otherwise turning centuries of science on its head, be prepared to climb a steep hill.

For our conduct, we want to stay respectful and constructive. We will investigte the concepts on grounds of theory and experiment, in an orderly way, moderated as necessary to stay on topic. All discussion should pertain to the concept at hand, and its technical aspects. And of course, we want to stay impersonal, since we are seeking the objective reality.

It is honorable to help one's colleagues find the flaws in their concepts, and it is equally honorable to seek the flaws in one's own concepts. Help yourself to understand if a breakthrough could finally be at hand!

As an editorial aside, my own opinion (Williams) is that it's too early to worry about funding. I know everyone has to eat. But funding must necessarily come after a believer has been made of the second person. So we will need day jobs til then, but that is not unlike any visionary who ever went before.

Today's space industry is different than the government-directed concepts for space exploration that have been nurtured since Apollo. Instead, it seems that a good idea will find a tech entrepreneur or venture capital firm. The big changes being made in the space industry today stem from that model, so perhaps breakthrough propulsion will as well.

Enough philosophizing. Let's get on to the workshop!

Estes Park Quick Study I
The Fuel Problem
one of two fundamental obstacles to interstellar travel

Travel between the stars, and even between planets around a star, requires propulsion to overcome the effects of gravity. A propulsion source drives motion of the spacecraft with respect to the nearby stars or planets, lifting it out of the local gravitational well.

Chemical rockets are the main propulsion source used in the interplanetary programs. They operate on the principle of throwing something out the back so that the spacecraft is thrown forward. To accelerate a chemical rocket at 1 g half way to the nearest star, and decelerate at 1 g the other half way, would require a fuel tank the size of the moon. Therefore, rockets achieve a speed limited by the fuel they can practically carry. The speed is so small as to require centuries to reach the nearest stars.

Other fuel options include ion beams, microwave sails, and reflecting the blast waves of nuclear explosives. These options do not seem to alter the basic feasibility of chemical rockets, because all these alternative methods yield velocities as small as those of chemical rockets.

To solve the fuel problem would be to provide a limitless source of fuel that could be used to power a spacecraft indefinitely. This would presumably be through some aspect of the universe that is everywhere existent.

It is understood that even with the fuel problem solved, the time-distance problem would remain. The fuel-problem would only allow the continuous acceleration of objects up to speeds approaching that of light, but a solution to the fuel problem alone would not be sufficient to achieve hyper-relativistic travel.

L.L. Williams
July 2016

ESTES PARK QUICK STUDY II

The Time-Distance Problem
one of two fundamental obstacles to interstellar travel

The Time-Distance Problem is the really profound problem of interstellar travel. Even if the fuel problem were solved, and we could accelerate freely in any gravitational field, accelerating even up to the speed of light, our civilization could still not explore the stars. This is because the severe effects of time dilation would isolate any emissary or probe in time, very much like a Planet-of-the-Apes scenario. Our astronauts could see the center of the galaxy, but they would return to the far future of their planet, and their civilation would be gone.

It is best to approach this problem mathematically, so that we can be awed by the fundamental simplicity of our obstacle. In essence, the time-distance problem is inherent to the structure of space and time.

We understand space and time to be joined together in a spacetime continuum. Moreover, "distance" between any two events in spacetime is invariant with respect to the state of motion.

$$c^2 d\tau^2 \equiv c^2 dt^2 - dx^2 - dy^2 - dz^2 \equiv \eta_{\mu\nu} dx^\mu dx^\nu \qquad (1)$$

where t is a time coordinate, x, y, and z are spatial coordinates, τ is called the proper time, and c is the speed of light. Greek superscripts denote the 4 components of space and time.

In terms of 4-velocity $U^\mu \equiv dx^\mu/d\tau$, the invariant interval (1) implies

$$\eta_{\mu\nu} U^\mu U^\nu = c^2 \qquad (2)$$

This equation implies

$$\eta_{\mu\nu} U^\mu \frac{dU^\nu}{d\tau} = 0 \qquad (3)$$

It's a general rule of motion in spacetime – we haven't done any physics yet – that the 4-velocity and the 4-acceleration are orthogonal vectors.

Consider a spaceship moving in one direction. In the rest frame of the moving spaceship, where the coordinate time is the proper time, $U^\mu_{ship} = (c, 0, 0, 0)$, and $U^2 = c^2$ in all frames, consistent with equation (2).

Consider now the simple case of constant acceleration in the x direction. One viable trajectory to the stars would be to accelerate at 1 g half way, and decelerate at 1 g the other half way; thereby, maintaining artifical gravity for the astronauts. From (3) we deduce that in the rest frame of the spaceship, $(dU^\mu/d\tau)_{ship} = (0, a, 0, 0)$, where a is the acceleration or effective gravity measured in the frame of the spaceship. It follows that $(dU/d\tau)^2 = -a^2$ in all frames.

Therefore we obtain the two equations in two unknowns $t(\tau)$ and $x(\tau)$:

$$c^2 = \eta_{\mu\nu} U^\mu U^\nu = c^2 \left(\frac{dt}{d\tau}\right)^2 - \left(\frac{dx}{d\tau}\right)^2 \tag{4}$$

$$-a^2 = \eta_{\mu\nu} \frac{dU^\mu}{d\tau} \frac{dU^\nu}{d\tau} = c^2 \left(\frac{d^2 t}{d\tau^2}\right)^2 - \left(\frac{d^2 x}{d\tau^2}\right)^2 \tag{5}$$

The solutions to this pair of equations are:

$$t(\tau) = \frac{c}{a} \sinh\left(\frac{a\tau}{c}\right) \tag{6}$$

$$x(\tau) = \frac{c^2}{a} \cosh\left(\frac{a\tau}{c}\right) \tag{7}$$

These simple equations tell us how time and distance pass for a ship under constant acceleration. The parameter τ is the time coordinate on board the ship. And we haven't even done any physics!

Equation (6) tells us that for acceleration a at 1 g, and onboard elapsed time τ of 5 years, 69 years would pass on earth, far longer than a typical space program lifecycle. Yet that would scarely allow the astronauts to reach the nearest stars.

Equations (6) and (7) can be combined to demonstrate the limiting speed of light:

$$\frac{dx}{dt} = \frac{at}{\sqrt{1 + a^2 t^2/c^2}} \leq c \tag{8}$$

The time-distance problem is that motion in spacetime introduces severe time dilation effects that completely cut off contact between the earth and any distant explorers. And we haven't even done any physics!

The warp drive solution to the Einstein equations by Alcubierre, and the worm hole solutions to the Einstein equations studied by Kip Thorne, would get around this speed limit. By bending spacetime itself, effective hyper-relativistic speeds can in principle be achieved; but not in practice. Not yet, anyway. Another way around the time-distance problem might involve hyperdimensions. But the time-distance problem is quite profound and fundamental.

L.L. Williams
July 2016

ESTES PARK QUICK STUDY III

Warp Drives and Wormholes: Good News, Bad News

The good news is that general relativity *appears* to allow, in principle, the traversal of distances faster than light speed, and therefore offers a solution to the Time-Distance Problem.

From equation (8) of the Time-Distance Problem summary (Quick Study II), you might infer that the limiting speed will be the coefficient of the time component in the metric. In flat space, $\eta_{tt} = c^2$. In the curved spacetime around a black hole, the Schwarzschild metric gives the time-time component as

$$g_{tt} = c^2 \left(1 - \frac{2GM}{rc^2}\right)$$

We conclude the speed of light gets smaller as one approaches a star, and note that the limiting speed c need not be the limit everywhere in spacetime. Since mass affects the speed of light, one might hope to control the shape of space and time to overcome the time-distance problem. This is where warp drives and wormholes come in.

The Einstein equations describe black holes, warp drives, and worm holes. They are written:

$$R_{\mu\nu} - \frac{1}{2} g_{\mu\nu} R = \frac{8\pi G}{c^4} T_{\mu\nu} \qquad (1)$$

These equations are in the 10 unknowns $g_{\mu\nu}$. The quantities in $R_{\mu\nu}$ and R are second order derivatives of the $g_{\mu\nu}$.

The quantity $T_{\mu\nu}$ is the stress energy tensor, with units of energy density. Some are surprised to learn that there is no formula or prescription for choosing $T_{\mu\nu}$, but there are many standard forms for various situations.

The Newtonian analogue of (1) is

$$\nabla^2 \phi = 4\pi G \rho \qquad (2)$$

In (1) and (2), a second derivative is related to a mass/energy density, with the coupling constant G. Therefore, curvature of space is quantified by the magnitude of G. The units of the LHS of (1) are $1/l^2$. Therefore (1) gives a curvature scale:

$$L_\rho \sim \sqrt{c^2/G\rho} \qquad (3)$$

We may reasonably hope to transcend the light barrier by building a wormhole or bubble of some sort, and (3) allows us to estimate how much mass density is needed for a given size warp or hole. A 15-meter bubble requires a mass density of 1 earth mass per cubic *meter*. This seems far beyond our means. Because the gravitational constant is so small, astronomically large amounts of mass density are needed to curve space appreciably.

If we substitute $\rho \sim M/L^3$ in (3), then we recover the Schwarzschild radius of $R_S \sim MG/c^2$. The Scwharzschild radius of the earth is 4 millimeters. Even cobbling together a mass the size of earth could only give you a warp or wormhole big enough to transport a flea.

But if that is not enough bad news, there is more.

The equation (1) universally describes an attractive force of gravity. Einstein did not anticipate any sign change on the RHS of (1). But to build a workable bubble or wormhole, you need to come out what you went in, like Jonah from the whale. Therefore we need the mass energy to push "out" like antigravity at some point. For this reason, a minus sign is optimistically inserted on the RHS of (1). This is sometimes called "negative energy".

The problem is we have never seen or measured a negative energy substance. Perhaps the best known example is the quantum vacuum, but the measured value of the vacuum energy in cosmology is orders of magnitude smaller than what we expect from quantum theory. So we really don't even understand the quantum vacuum, let alone other forms of exotic energy.

The takeaway is that wormholes and warp drives are only achievable with astronomical quantities of something as yet unknown to science. And that's bad news.

L.L. Williams
July 2016

| Fuel Problem Time-Distance Problem | ✔ Theory ✔ Experiment | ✔ Existing Physics ✔ New Physics |

General

- Lance Williams will lead a short discussion regarding acceptable extensions to existing laws of physics
- George Hathaway will lead a full block discussion on aspects of valid experiments

Issue Summary:

To start the conference, we wanted to spend some time giving an overview of expectations for a valid theory or valid experiment. We have made a lot of progress with the scientific method, and we feel sure we are not going to get to the stars without it. That means repeatable, verifiable experiments; and theories that conform to expectations from modern physics.

Lance will use a short block to present a summary of the aspects of a valid extension to the laws of physics. These aspects are few, but profound. These aspects incorporate the criteria that Robert Dicke used when he set about verifying general relativity versus other possible theories of gravity.

George will lead a discussion on the aspects of a valid experiment, cutting across theories and effects of all types. George is particularly interested in low-thrust experiments, and so this is quite relevant to the Mach Effect and RF cavity thruster sessions to follow.

These two preparatory sessions should lay the groundwork somewhat for the type of scrutiny on theory or experiment to expect in the subsequent sessions, and which will be captured in the proceedings.

Aspects of Plausible Extensions to Physical Law

L. L. Williams
Konfluence Research Institute
Manitou Springs, Colorado, USA

WE DISCUSS SOME OF THE PROFOUND CONSTRAINTS ON ANY VIABLE EXTENSION TO THE KNOWN LAWS OF PHYSICS.

1. Theory Without Experiment

We can anticipate two basic situations in our search for a propulsion breakthrough. One is a compelling experiment with no theory. If someone can reliably produce an effect to levitate a cannonball, then the discovery is at hand and the theory can be developed from observation of the effect. More likely is the second case, in which we seek experimental confirmation for proposed modifications to physical law. It is for this second situation that we consider some aspects of allowable extensions to the laws of physics.

As we are likely considering extensions to the laws of gravity, we adopt the framework that Robert Dicke used in the 1950s to consider theories of gravity alternative to general relativity.[1] These alternatives could be parameterized against observation to verify general relativity against other plausible theories.

2. Covariance and Lorentz Invariance

The first constraint is that the theory must be covariant: it must keep its form under coordinate transformations. Furthermore, the space and time coordinates must satisfy the Lorentz transformation. In practical mathematical terms, this means the equations must be written in terms of 4-vectors or tensors with 4 degrees of freedom per index. All current laws of physics, classical and quantum, are covariant.

Here are some examples of covariant or Lorentz-invariant equations:

$$R_{\mu\nu} - \frac{1}{2}g_{\mu\nu}R = \frac{8\pi G}{c^4}T_{\mu\nu} \quad , \quad \frac{dU^\mu}{d\tau} + \Gamma^\mu_{\alpha\beta}U^\alpha U^\beta = 0$$

$$\partial_\nu F^{\nu\mu} = \frac{4\pi}{c} J^\mu \quad , \quad \frac{dU^\mu}{d\tau} = \frac{q}{mc} F^{\mu\nu} U_\nu$$

By way of comparison, here are some equations that are generally not covariant or Lorentz invariant:

$$\mathbf{F} = \mathbf{E} \times \mathbf{B} \quad , \quad \nabla^2 \phi = 4\pi\rho \quad , \quad \mathbf{F} = m\frac{d\mathbf{v}}{dt}$$

The non-covariant forms above are encountered in specific coordinate systems, and physicists work with them every day. But they cannot be the starting point. Forms like those above must follow by writing a covariant equation in a specific coordinate system.

3. Lagrangian

A second important constraint from the Dicke framework is that the theory have a Lagrangian. All known laws of physics, classical and quantum, are derivable from a Lagrangian. The Lagrangian generates the field equations and equations of motion according to a fixed operation. There is no method for finding a Lagrangian; it must be guessed, its equations derived, and its predictions checked against experiment. Weinberg won the Nobel Prize for a 2-page paper guessing the correct electroweak Lagrangian.[2]

The Lagrangian for the known classical fields of gravity and electromagnetism is:

$$\mathscr{L} = A_\mu J^\mu - \frac{1}{4\mu_0} g^{\alpha\mu} g^{\beta\nu} F_{\alpha\beta} F_{\mu\nu} + \frac{c^4}{16\pi G} g^{\alpha\beta} R_{\alpha\beta} + T_{\mu\nu} g^{\mu\nu}$$

where A^μ is the electromagnetic vector potential 4-vector, J^μ is the electric current 4-vector, $g_{\mu\nu}$ is the gravitational metric tensor, $F_{\mu\nu}$ is the electromagnetic field strenth tensor, and $R_{\mu\nu}$ is the Ricci tensor.

It is remarkable that all the complexity of gravity and electromagnetism, with their tensor field equations, is captured by a scalar entity. An enormous amount of information is unpacked from the simple package of the Lagrangian.

I have not bothered with the quantum Lagrangian of the strong and electroweak forces. My own view is that reaching the stars is a classical, not quantum, problem.

Rodal: I agree the Lagrangian is important, but the Second Law of Thermodynamics cannot be written in terms of a Lagrangian. Some of the laws and effects we are dealing with involve dissipation, which cannot be written in terms of a Lagrangian. More than a Lagrangian may be needed for our case.

Williams: I take your point. The laws of thermodynamics are important and have no Lagrangian. However, we are focusing here on particles and fields, and the laws of thermodynamics are not specific to any particles or fields. They are general properties of natural processes.

4. Falsifiable Prediction

A third constraint is to demand a concrete prediction. After all, there is no point to the theory if it makes no prediction. Closely related to this is that an experiment exists to prove or disprove (falsify) the theory.

A new viewpoint, but without a new prediction, is not a new theory. Feynman famously developed several alternate viewpoints to help him understand the particular theory under development.

And that concludes my short summary of some aspects of legitimate extensions to the laws of physics.

5. References

1. see section 2.1 of *Theory and Experiment in Gravitational Physics*, by C.M. Will, revised edition, Cambridge University Press, 1993.

2. S. Weinberg, *Phys. Rev. Lett.*, **19**, 1264, 1967.

Discussion

Robertson: Give me an example of a new prediction.

Williams: One example which I will give later is a tunable coupling of gravity. If there is a coupling between gravity and electromagnetism, and I can do some electrical experiment that changes the weight of a cannonball – that's not in existing physics. So a new prediction is an effect not known to physics and not contained in current physical law. Another example is Planck's evaluation of the blackbody spectrum and its resolution of the ultraviolet catastrophe. That was something new to physics at that time. It should generally manifest ultimately as a new experiment not explainable in existing physics. However, Jim's work is an example of a case in which the new effect is in existing physics.

P. Jansson: You agree that string theory is not falsifiable. So we have a higher standard on theories here than is demanded of mainline string theory.

Williams: That's a good point.

Turner: Doesn't a new theory need to fit into the old theory in certain limits? For example, Newtonian theory is found in general relativity in the limit of small velocities, etc. I thought that is what Tony was saying.

Williams: Yes that's a good point. Perhaps I should have listed it as another constraint, that it fit into the broader framework of known physics, and not violate or contradict anything there.

Tajmar: String theory can actually predict a violation of the equivalence principle. A space mission is being prepared to test that effect. So string theory is falsifiable.

Hathaway: Last night after dinner, we heard Brandenburg's prediction, or calculation, of the mass of the proton. Would you consider that a new prediction? In that case, there is no experiment, yet it is a new prediction.

Williams: That's a good point. An example of that type might be Bohr's calculation of the Rydberg constant.

Christie: I have a sort of foundational question. Why are space and time connected? Is anyone working on why that is?

Williams: It's essentially built into general relativity. The connection between space and time is what we call gravity. But it has a sound empirical basis going back to the Maxwell equations.

group discussion of the constancy of the speed of light...see video

Robertson: I want to make an argument against Einstein's general relativity. If general relativity had come after quantum physics, then general relativity would be consistent with quantum physics and we wouldn't be having this discussion. Isn't our obeisance to general relativity in part an accident of history?

Brandenburg: It has proven impossible to reconcile general relativity and quantum theory

Hansen: We should be careful to not disregard quantum mechanics. It may hold something important for the breakthrough propulsion problem.

Experimenting With Novel Propulsion Ideas

George Hathaway
Hathaway Consulting Services
Toronto, Canada

About Myself

I am a professional engineer, graduated EE in 1974 from the University of Toronto. I own a little company called Hathaway Consulting Services (HCS) near Toronto, Canada. HCS was established in 1979 and has an international clientele: foundations, private investors, institutions & agencies. The focus is on exotic technology: primary areas being propulsion, gravity, energy, and materials. The primary mandate of HCS is fundamental experimental research. I appreciate being invited to this workshop; it's an honor and I'm looking forward to all the talks.

1. Introduction: HCS Capabilities

The bulk of my talk is going to be about measurement pitfalls and prosaic explanations for what is seen in the lab. However, I will give you a little overview to begin with about what we do. HCS has been established since 1979. It is a private organization. We are not associated with any government agency. We are not funded by any agency or institution. It is primarily private investors, private clientele, private foundations in North America and Europe.

Our operation is basically twofold. We provide a service for inventors: people who have what is considered a crazy enough idea that it might just be worthwhile looking into and funding. So, I will try to attract funding for inventors. We also, on the other hand, have a service that we provide for investors. Often a venture capitalist for instance, as they do quite often, has some group or person come up to them saying, "I've got the answer to space propulsion," or, "I've got the answer to free energy," or something like that and they don't know where to go. These are the areas in which I particularly specialize. A university typically won't touch it. It's not within the paradigm of what Lance has so adequately described just before my talk. Even the DoD or DARPA might say, "That's a little too flaky even for us." Then, if the investor knows about me, I will look at the invention and we will provide that service, primarily from an experimental standpoint, but also we have theorists that we can call on; for instance, the Institute for Advanced Studies in Austin, Texas.

Our capabilities have grown substantially since our inception and I won't go through all of them. But the reason I'm listing some items is in case some of you folks need a particular experimental capability that would be useful to you in one of your experiments. The list of lab capabilities below is a small subset of what is available. The lab is about 11,000 square feet of space with all sorts of wonderful and bizarre things. We not only can stimulate the experiment but we can measure the response with analytical instruments. We have been privileged also to produce some of the piezoelectric transducer crystals that Jim has used or was going to use in one of his experiments. We have a material science lab too where we produce magneto- and electro-active ceramics as well as other specialized materials. Here is a partial list of HCS capabilities:

1. cryogenic liquids and gases (to liquid He temperatures)
2. high magnetic fields, both pulsed and DC
3. microwave hardware, waveguide & cavity design
4. ultrahigh pressures and temperatures for novel materials processing
5. high voltage (up to 600 kV) ultrafast (sub–nS) pulsers and radiators
6. electric arc–induced shockwave studies in liquids
7. sensitive vacuum balances for gravity modulation & manipulation studies
8. RF anechoic chamber; large & small high vacuum chambers
9. unique apparatus for materials fabrication and testing
10. high–temperature ceramic superconductor manufacture
11. development of novel ultra–wideband ferrite, piezoelectric & dielectric materials
12. high–power RF designs both solid state and vacuum tube
13. test beds for gyroscopic and other mechanical and electrical thrusters
14. design and testing of apparatus to investigate energy production from quantum vacuum
15. wet lab for bio-communication research
16. SEM/EDAX, TEM, Confocal, RAMAN, XRD, Mass Spec, EPR/NMR, materials testing, high-speed cameras, gravimeter

Rodal: Do you have a sintering press?

Hathaway: Yes, we have various vacuum and controlled-atmosphere sintering furnaces and Cold and Hot Isostatic Presses. We have a 600 ton uni-axial press with a large die we made for the Podkletnov spinning disk experiment the results of which I published in Physica C in 2003. There we made 6 inch high temperature multi-layer YBCO superconductors.

Rodal: Did you grow the crystals? How large did they get?

Hathaway: Yes, probably on the order several centimeter but growing crystals is something we haven't done for a while but, let us know if you need some crystals.

Cole: What are the gravity meters you used, are any of them commercial gravity meters?

Hathaway: Yes, the one I'm listing here is not automated. It's sensitivity is about one part in 10^8g.

Cole: We used one before and the resolution was about an hour, not even seconds. So, it was basically worthless.

Hathaway: For high-speed studies, no. Ours uses a quartz gravimeter with a few seconds time constant. It has a tiny quartz lever with a little platinum weight on one end and you peer into it. It's primarily a commercial device used for geophysical surveying. But, it's available so if you have a long duration experiment that isn't a pulsed experiment, it can still be very useful. And we have balances that do react faster but don't have the sensitivity.

2. HCS Propulsion Experiments

Some of the propulsion experiments that we've been involved with, either designing or testing for various clients, are listed here. This is only a subset of some of the general studies we do in the lab which, as I mentioned are generally propulsion and energy. We've had a long history of involvement with people who have come to us with new energy devices. The list regarding energy is much longer than what is shown below. This is a propulsion workshop so I've emphasized the propulsion side of things.

Below is a partial list of HCS propulsion experiments. We have investigated Biefeld–Brown HV capacitors, John Brandenburg's GEM theory-based rotating currents and Barrett/Froning SU(2) coils, all with null results. These will not be further discussed here. A few that I will highlight in this talk are:

1. Graneau water-arc discharge thruster
2. gyroscopic force rectification
3. Williams div(J) & gravity

4. Podkletnov rotating superconductors & gravity beam

5. Zinsser HF force accumulation

6. spin-polarized nuclei/gravity interaction

7. Hutchison Effect

8. Woodward, Mach Effect drives

There are many individual experiments we've done on rectification of rotational motion to linear force, which is an old standby, and a lot of these particular experiments deal with purely mechanical devices. An inventor might say, "I've got a gyroscope and I put it on a string and it sort of swings over this way a couple of times and its average looks like it has a thrust in a certain direction". But, when you actually do the proper experiment, the average thrust is zero.

Cole: I don't want to criticize any experiments that anyone is doing in here, but I will criticize how it's being publicized. The publications a lot of these experiments I read and the impression I get from reading them is that what that person said would happen is untrue, when really what the experiment has actually done is a subset of what the person did, not the true experiment of what the person did. Yet, the article reads, "What that person did is wrong." And so you need to bring up when you're writing these papers is that you're in some kind of limit of the original experiment and not doing the original experiment, but that doesn't seem to be brought out in these papers. It's almost like you're saying we did the original experiment and it didn't work when you really did not replicate the original experiment.

Hathaway: Yes, for instance my Podkletnov reproduction was criticized by saying that I had not done the actual experiment. I had not said this is not a true representation. There are gradations as you pointed out. It is not always possible to perform an exact replication of an experiment which has already been designed or carried out or for a theory that has been developed and tested. Unfortunately people don't have a long enough attention span, or the ability to read. Anyway, we can come back to that.

1. Graneau – energy from water by H–bond breaking

Some of you might have heard about Peter Graneau and his son Neil. We spent many years with the Graneaus, specifically in their energy experiments. They were suggesting that if you introduce a high current discharge from a capacitor bank very quickly into water, the water would produce around the arc a huge quantity of "fast fog" they called it. The

Figure 1: Peter Graneau shown with his high current discharge into water experiment.
Fast fog from a cannon shot a projectile up to a catcher box above.

fog would shoot up through the water into the air above and if you calculate the momentum and the energy in the fog in the air, it would be greater than the energy introduced by the capacitor bank in the first place [1]. Hence, there was some weird over-unity energy activity going on which they ascribed to something called hydrogen bond breaking in the water. That was an energy experiment, but it wasn't generally known that we also considered using this idea, whether it was over unity or not, as a propulsion system. So, we did some experiments. Here you can see (see Fig. 1) a high voltage pulsed power supply. There's an arc discharge device, a little water cannon down there, that is going to shoot fast fog which certainly is fast when it's coming out of the barrel and we're timing it as it pushes a light projectile up into this little catcher above.

This would have been a very interesting propulsion system even though it was classical. We would push pulsed fog out the back to provide forward momentum. But it would have been much more exciting if there was an over-unity component to it. Unfortunately, there was not.

2. Gyroscopic motion levitation experiment

Here is an example of what would be considered the grand-daddy of gyroscopic precession propulsion systems. This device stands almost a foot tall. (see Fig. 2) The rotor is on the order of 7 or 8 inches in diameter. The hoped-for outcome was that if we precess a spinning gyro at the correct ratio of the nutation frequency compared to the rotor spin frequency, we might get this thing to lift off, or at least lose weight.

That's an old thought that has not proven itself in any experiment that I've ever known or been involved with. But I was involved with this one and these guys really went to town with the force gauges underneath and all sorts of instrumentation all over the place and they were just getting a whole bunch of noise (see Fig. 3). The investors put millions of dollars of high tech into this experiment.

Figure 2: Gyroscope being testing for weight loss

Figure 3: Gyro being tested at HCS.

So, I made my own little reproduction (Fig. 3.) with a rate gyro from a WWII aircraft, doing basically the same as that experiment. But theirs was tightly instrumented, and it was not actually allowed to move much. So we experimented with several ways of seeing whether there was a force generated. One was on the end of an optical table with the device on the end of a simple swivel arm, and it would only oscillate back and forth. (see Fig.4) It never progressed its oscillation forward. Another simple way of testing for constant thrusts is a ball table (see Fig. 5). A lot of people denigrate this method but it's actually quite sensitive. The grey area in Fig. 5 is a granite machinist's table which is ground flat within tenths of a mil. There's a thick glass plate on which the experiment sits and there are plastic (or steel) balls underneath. On the glass plate is a little white card with a cross in the center. A machinist's height gauge with pointed tip on an arm is placed on the table with the pointed tip aimed down onto the center of the cross. If there is a net thrust, the cross should move away from under the tip. My experiment just vibrated and jiggled around but the cross on the card under the pin showed no progressive motion.

3. Williams' 5–Dimensional Theory

Pharis Williams's 5-dimensional theory suggests you can produce a region of reduced gravity between two conducting plates on which high current

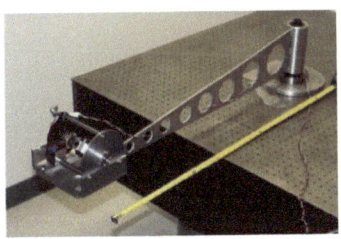

Figure 4: Gyro tested at HCS, using pivot arm.

Figure 5: Gyro tested at HCS, using ball table.

is respectively diverging and converging from the periphery to the center (div J). In the experiment, you produce a strong divergent current, which diverges from a point on one conducting plate and on a second plate nearby, the current is converging to a point. Then, according to his theoretical calculations, there should be a change in gravity between the plates. We devised an experiment for Williams where we had a high-current conducting rod attached to the center of the flat end of 4" copper plumbing end cap and another rod attached to the center of a second end cap which was facing the first and very close to it. A thin dielectric disk was suspended between them whose weight was measured during current conduction. (see Fig. 6). The 2 rods were attached via welding cable to a battery bank of several thousand amps through a high-current contactor. The circuit was closed by the O-shaped copper plumbing.

Unfortunately the dielectric disk didn't change weight even though we were able to measure the weight between the diverging and converging plates down to a factor much better than he was expecting.

Meholic: How did you measure the weight of the central dielectric plate?

Hathaway: The whole copper O-shaped apparatus was placed in a frame as shown but what you cannot see is a hole in the top left elbow. A thread supporting the dielectric disk passed through this hole and was attached to a sensitive analytical chemical balance above the apparatus. The top high-current rod was hollow to allow the thread to pass through.

Our experiment did not say that there is no anomalous force or a gravitational interaction because of diverging and converging currents according to Williams's theory. It just means that we were not able to detect it down to level that he wanted.

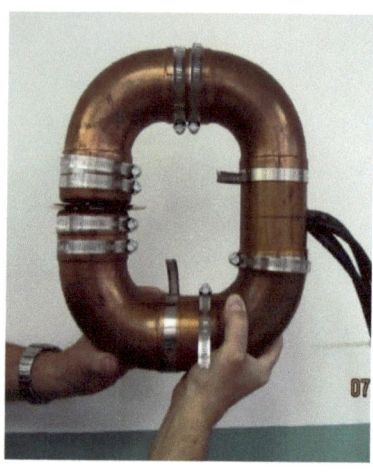

Figure 6: Williams' 5–Dimensional Theory Test Apparatus.

4. Podkletnov Spinning Superconducting Disks

In Fig. 7, there is Eugene Podkletnov (lower right) on one of his two visits to our lab in 1996. We are looking at an experiment that we reproduced from an experiment that had been done at the University of Turin in 1992 I think. There, a small Yttrium Barium Copper Oxide (YBCO) superconductor was spinning in the vapors of liquid nitrogen. It was in the Meissner state, but their method of looking at weight loss was not very sophisticated. The Turin setup was at the undergraduate level to see if any anomalous effect was present but our version had quite a bit more precision. We did not see the weight change. There were some strange transient effects when you start the spin up or you slow down suddenly a superconductor, but we could never ascribe them to anything other than instrumentation noise and spurious thermal effects.

We went on to make the larger experiment, the reproduction of Podkletnov's spinning superconductor experiment. The guts of it are shown in Fig. 8, which is an insert for a large liquid helium cryostat. There are 3 solenoidal levitation coils, two of which you see prominently at the bottom, copper colored. In this experiment there are also 3 high-frequency (5 MHz) coils which loop through the central hole in the superconductor. These are seen just above the levitation coils. And on top, the 2 plates of aluminum which hold the gearing mechanism I used to spin the superconductor even though, topologically, it also had these loops of wire running around it. So that was a nice little design challenge.

Our paper was published in Physica C in 2003, [2]. I'm convinced Podkletnov never actually went to all this trouble.

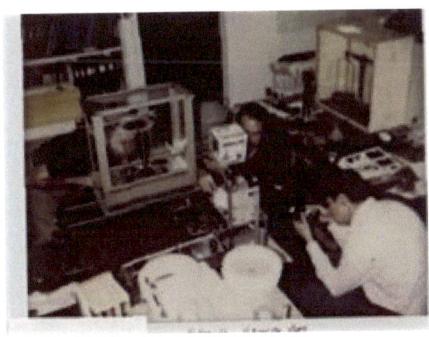
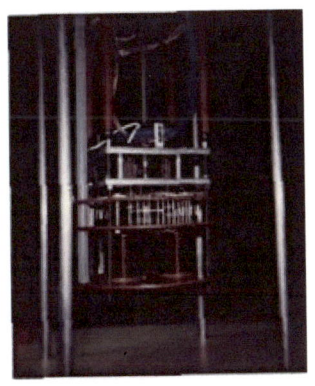

Figure 7: Podkletnov, small table-top spinning superconducting disks.

Figure 8: Podkletnov large disks setup.

Williams: So those were all null results?

Hathaway: It was null to within the measurement resolution of our equipment. Once again, I cannot say there is no effect. All I can say is that to the best of our ability, which was approximately fifty times better than what he had claimed, we saw a null result.

I suppose most people in the room have also heard about his so-called gravity beam experiment [3]. Podkletnov claimed that at high enough DC impulse voltage applied to a high-temperature superconductor, a flash of something magical will boil off the superconductor and head towards a target, in this case a grounded copper ring. Some magical beam of gravitational force will emanate from the other side of this ring and travel through space to impart a ponderable force on objects in its path.

Figure 9: Podkletnov gravity beam experiment.

I learned about that experiment when I went over to Europe to attend a lecture that he gave on the results of his initial beam experiments using a van de Graaff machine. I don't think anyone in this room was at that lecture back in the 90's when he first announced he had done that experiment. A few years later we had built two 600 kV van de Graaff machines for a different experiment. Using these, I put together a Podkletnov gravity beam experiment, where the superconductor is the black thing you see inside the horizontal glass vacuum tube. It's glued thermally to a little liquid nitrogen holder, and there's a little liquid nitrogen dewar inside the left inner dome here. The target is a grounded copper disk. The whole thing is pumped down using a turbo pump vacuum system.

We "aimed" this into the Faraday cage or screen room shown behind the apparatus, and we used a very sensitive force detector in that cage, and that was a wire "clothes hanger" with strips of toilet paper hanging from it. It turns out, that is extremely sensitive to small forces, much smaller than the forces claimed by Podkletnov to, for instance, knock over objects on a desk. So we were video taping the hanging toilet paper and operating this gravity beam rig. We got a null result.

Brandenburg: Is the superconductor actually a dual–layered superconductor?

Hathaway: This was actually a Murakami-style melt-textured (polycrystalline) superconductor we made in-house, because Podkletnov had used melt-textured in one of his very first gravity beam experiments. This is a three inch diameter melt-textured YBCO superconductor.

Brandenburg: But when he did his experiment he used a double–layer superconductor.

Hathaway: He did also use a double layer in some experiments.

Brandenburg: The rotating ones, were basically I think just crushed superconductor...

Hathaway: He actually did use three-layer disks, too. Note for some spinning disk experiments he specified sintering YBCO, then crushing and sieving, then sintering again to get the desired grain size distribution. We did make a three-layer to his specifications, which was two superconductors sandwiched with a praseodymium layer to kill the superconductor as the middle layer. He had all sorts of different kinds of materials he suggested would work.

Brandenburg: I have a little aside comment here. There was one particular expedition to the Livermore Lab, where we were tasked to try to reproduce a Russian result and actually got it work. In this case it was an experiment done by some very reputable Russian scientists. They published, as usual this is the Cold War, a very terse article describing vaguely how they got these marvelous results. And we actually got the thing to work to their amazement. We concluded later that the whole thing was a wild goose chase. We were doing laser pellets in one dimension, you know, everything converges absolutely spherically in one dimension. If you throw in two-dimensional effects everything goes all over the place, it's like scrambled eggs. But in one dimension we could get their stuff to work, and they had left out so much stuff that we kind of just put in my guesswork. I'm just saying that the Russians do send people on wild goose chases and part of it is to knock the system over here. They want to see what the reaction is, and also, it's to sop up money, any money that is actually going to worthy causes so – pardon me.

Hathaway: They're also interested in knowing what our technical and analytical capabilities are. How far are we advanced in the ability to measure these things?

5. Zinsser impulse accumulator based experiments

Probably not many people know a German experimenter named Rudolph Zinsser [4] who had a theory and an experiment where he claimed that if you produce 40 MHz saw-tooth waves, and introduce them into water in a certain way, and weigh the water, you will get an accumulation of force impulses. We experimented with this idea. Zinsser had explained this experiment and demonstrated this effect at a conference I held in 1981 at the University of Toronto. He had shown that on his balance that he brought over from Germany, he was able to have this container full of water and 2 electrodes actually lose weight as these force impulses accumulated in the water.

So, many years later, in 2003, I got around to finally doing the experiment properly. And properly means having water in a vacuum vessel. Zinsser did all his experimentation in air. The water has to be contained in a water-tight, vacuum-tight vessel. For the RF sawtooth energy to flow into the water, it has to go through the vertical capacitor plates shown on top of the water container which allow the vertical movement of this vessel on a balance beam without wire connections. That balance beam is inside a tube exiting the far side of the large vacuum chamber shown in Fig. 11. I used one of Jim Woodward's optical displacement sensors which he kindly provided to me some time ago. Thank you very much Jim, I appreciate it. Sadly, null results ensued.

6. Nuclear spins and Gravity

We were also involved with experiments to determine relationships between nuclear spin polarization and gravity. Here's an experiment that involves electron parametric resonance (EPR) which aligns electron spins which will then, by the Overhauser Effect (also called DNP - Dynamic

Figure 10: Zinsser Experiment: water plus electrodes in acrylic container.

Figure 11: Zinsser experimental vacuum chamber..

Nuclear Polarization) align the nuclear spins in a much more effective way than simply by nuclear magnetic resonance (NMR) alignment. When the electron spin system is in thermal equilibrium, the polarization transfer from electrons to nuclei requires continuous microwave irradiation at a frequency close to the corresponding EPR frequency. So we require a microwave cavity and associated hardware and a very sensitive vacuum balance and vacuum system and a sample which is cryogenically cooled. So far this experiment is still underway and there are no results yet.

Figure 12: Nuclear spin and gravity connection.

7. John Hutchinson

We also tested something called "The Hutchinson Effect". John Hutchison had a famous video of a cannon ball levitation, amongst many other bizarre occurrences.

I bring up John Hutchison because of what Lance had mentioned earlier, namely you can have experiments and no theory. There's no theory to explain this guy or any of the stuff associated with him. I've had lots of questions before the talk started here, like what do you think? Is Hutchison for real? All I can say, in the very brief time that I have, Hutchison was real at the time of the events that are described in my book [5]. Hutchison is not real now.

Mathes: What do you mean? He has moved onto the complex plane or what?

....audience laughter....

Hathaway: Not yet! Hutchison was able to levitate and break apart material and cause all sorts of other weird things to happen, which are described in my book, at the time that we were researching him. We had a contract to find out what was going on with this guy, and we tried our best but we were never able to discover what was going on. But, we witnessed and experienced the most unusual things that I have ever experienced in all my years being at this game.

Mathes: So you couldn't replicate it?

Hathaway: We couldn't but we set up the experiment ourselves, and we can go into this offline, but we set up the experiment ourselves just the way he did in a different location and we got some electrostatic effects. But only when he came and performed the experiments was there actual levitation, things like that.

Meholic: So on an independently constructed apparatus, he was able to operate it and get the results?

Hathaway: Yes, yes. The reason I'm bringing this up is that we are all talking about the technical aspects of advanced propulsion. You know, we've got theories and we've got quantum mechanics and we've got relativity and we've got experiments and all the stuff that I'm testing and that you guys have gone through. I just want to put in the very back of your mind the fact that consciousness might play a role.

8. Jim Woodward's Mach Effect Drive

Finally, we get to Jim's device [6]. I've seen very small effects that I believe are just above the noise in my thrust balance [7] using one of

Figure 13: John Hutchison in 1997 in Vancouver.

Jim's older first or second generation devices, a small PZT 19mm device. Fig 14 shows a picture of the vacuum chamber I built for the test and the associated electronics. What I saw was at 200V AC in magnetic shield in vacuum.

Figure 14: Vacuum chamber for Jim's test article HCS.

Figure 15: Balance beam with Jim's test article inside.

Woodward: It's a device that's the same as the one that Nembo tested and it's similar to the one in the setup we have now. Of course that device has been changed as you know, the brass reaction mass was changed as it produced much better effects. Did you get the new brass mass George?

Hathaway: Yes it's sitting there on the table.

Woodward: Aha! No you're fine. We're distributing new reaction masses to George and Nembo. We've produced a new device, and those are the reaction masses that should make it possible to see some slightly bigger effects.

A lot of our experiments have to be done on anti-seismic tables. The balance that we're testing Jim's and other thrusters on (and I'm sure Nembo will talk about as well) have to be free of seismic influences from the environment. It can never be totally free because the absorbing nature of, say, pneumatic bladders and such things do not get rid of all noise. But at least you have to know the frequency spectrum of the seismic vibration absorbing material so you can say ok, I'm working within this band of the noise spectrum which is outside the natural frequency of the absorber and you can then justify a little better than in fact you have removed that prosaic influence to a large extent from your experiments.

And it becomes a problem because anti-vibration damping does have frequency response. You can hit a resonance with some of the experiments and you'll get a false output. A lot of vibrations especially from vacuum systems become a real problem. If you are lucky, you can design a vacuum system that does not need hoses. The best way that we have found to get rid of tubing is ion pumps. Then, instead of having tubes or pipes that come from the vacuum system to your experimental chamber you now put these little ion pumps on. First, you still have to rough down the vacuum system but then you physically turn a valve off, and take off the hose, and connect the ion pump, which just has these little high voltage wires which are much less problematic coming down to your chamber than these big hoses.

3. Testing Issues Under the Assumption of a Completely Novel Invention

When we consider a test campaign, several general issues need to be addressed prior to designing the experiments when the assumption is that the invention is completely novel. Some are listed below. Somewhat different issues arise when contemplating a series of validation tests on an invention that has already been built, e.g., in areas of hypothesis testing.

- who is the test for: inventor or investor?
- design of suitable test bed for each project;
  costs of new equipment vs re-use of existing equipment
- cost/benefit of simple "look–see" experiments without full testing
- hypothesis generation (observed physical phenomenon without prior theory) vs hypothesis testing (confirmation/denial of prior theory)

- test protocol development: replication vs reproduction
- enumeration of likely prosaic/artifactual explanations
- control experiments
- statistical & error analysis
- instrument calibration
- minimum resolvable thrust required to prove claims

4. Introduction to Testing Nightmares

By way of introductory examples, an area in the list below is the effect of local gravitational variations where you have really sensitive experimental instruments. In fact, down in the low nano-Newton range that starts to have a significant effect. This one we ran into at Hal Putoff's lab in Austin. They were doing an experiment with a sensitive Cavendish balance and couldn't quite figure out why they could never zero it. One week they would be able to zero it at a particular rotation position of the torsion fiber and the next week they would have to turn it around and it turned out that somebody had moved a storage cabinet from one area to the other. It was about 20 feet away and it affected the zero position of their Cavendish balance. You would not think that would have any effect, but actually when they calculated it, the sensitivity of the balance was such that it did have this slight effect, this slight movement that was certainly enough to affect their final result.

Another example: Virtually all sensitive experiments that I know of have to be done under a vacuum of some degree depending on how you want to characterize them. A lot of stuff happens if you're not careful with how your vacuum system is constructed: where the ports are compared to your movable apparatus, the pumping rates, molecular drift within the chamber during the experiment, etc. You say, well, we pumped our chamber down and it was stable, at least from the gauge we put on the side of it. Either a thermal couple gauge or an ion gauge might have been attached to the side of the chamber and it has a steady reading so there are no transient pressure effects measured. In fact, you really don't know what's going on inside the chamber. The gases could be stratifying over time, causing an unstable, anisotropic situation versus what you think is a stable situation according to your gauge. All you have to look at is a gauge: a meter here or an ion gauge there saying we're down at 3×10^{-6} torr and it's been that way for half an hour. Well, in fact, inside there might be a whole bunch of other stuff going on gas wise, molecular flow-wise, that will take a longer time to settle down. That's just one component of the vacuum-related pitfalls.

Yet another example that comes up a lot is when you have a horizontal teeter-totter balance arm that is measuring something on one end, and you have a counterweight on the other end. A lot of "backyard" experimenters, at least those that are performing propulsion or thrust measurements, may have their thrust-producing device and/or counterweight rigidly fixed to and hanging at, let's say, 90 degrees from the balance arm. So, they see maybe an anomalous thrust happening with their balance but they forget that in fact the lever arm length is changing because the thruster or counterweight is a fixed angle. So, what they should be doing is pivoting the thruster so that you don't change the lever arm length.

Now we will overview the general experimental nightmares in the art of measuring: Are we seeing a real force, or is it some artifact of the experimental method? What follows is a list of effects which must be considered before claiming a new and game-changing result. A wise experimenter will make a check list and check off (and publish) each artifact as they are systematically eliminated from their experiment.

Question: Are you proposing, George, that everyone itemize and make sure that none of these artifacts are effecting their experiments?

Hathaway: No, I'm not proposing that they do anything. It's up to them to accept, or least be knowledgeable, about the fact that there are a myriad of effects that will interfere with their results. I'd be delighted if that was the case, but that will make an experimental report in some peer review journal excessively long. Usually, when someone really wants to get in and reproduce the experiment, they will go to the experimenter and say did you do this and this and run that test? So I'm not suggesting that one has to go through ALL of these items. Some items on this list are clearly not relevant to a particular experiment and can be ignored.

Woodward: I agree. You can't publish every test you did, but you have to be prepared when someone comes to criticize your work, to be able to answer their criticism that you did run though the needed precautionary tests and the result you are reporting is real and not some artifact.

Hathaway: If I were to write a check list of possible artifacts, then this is a subset of that list. It is by no means comprehensive, but covers a good subset of artifacts an experimenter should be aware of in this business.

NIGHTMARES IN THE ART OF MEASURING

I. Mechanical Effects

II. Temporal Effects

III. Electromagnetic Effects

IV. Electrostatic and Related Effects

V. Instrumentation Issues

VI. Signal Analysis

VII. Use of Controls

I. Mechanical Effects:

A. Thermal

1. Thermally-driven convective air or gas movements causing test masses connected to balances to move. Also, results from condensation of water vapor onto test mass or suspension during cryogenic experiments.

2. Radiometer effects on a test mass (in a radiometer, blade movement is caused by pressure of thin gas layers near blades due to absorption of solar energy).

3. Change of heat transfer conditions between test mass surface and liquid. This depends on i) delta-T between them which can change substantially over time, ii) thermal diffusivity of test mass.

4. Thermally-driven convective movement in liquid (usually cryogens) causing weight artifacts in submerged test masses connected to balances.

5. Change in length of lever arms or period of torsion balance due to thermal contraction/expansion.

6. Change in response of balances due to differential thermal expansion coefficients.

7. Short- or long-term temperature-induced drift of electronics in recording devices,
amplifiers/signal conditioners.

8. Thermal noise in balance structures, eg, torsion fibre and masses in a Cavendish balance.

9. Thermal gradients, and their time excursions, induced in test masses, especially superconductors (with corresponding distributions of superconducting or non-superconducting phases), resulting in only partial conditioning (e.g., only part of the superconductor is in superconducting state) due to insufficient or inefficient cool-down or

warm-up. Effect exacerbated by non-uniform test mass composition, density and thermal diffusivity.

10. Altered buoyancy of test masses, especially superconductors, in liquids (usually cryogens) due to free convection or 2-phase flow (gas bubble/liquid) in thin liquid layers close to the mass surface causing variations in liquid/solid friction.

B. Buoyancy

1. Different-shaped test, counterweight and dummy masses exhibit different buoyancy effects even in low-pressure gas.

2. Expected or calculated buoyancy of test mass or counterweight mass is enhanced or decreased by horizontal thermal stratification of still gas/air.

3. Account for buoyancy differences due to temperature differences even in low pressure gas.

4. Thermal shrinkage of test masses and supporting structures during cool-down causing reduction of buoyancy, e.g., in sample holders with large thermal expansion coefficients.

5. Absorption of water vapor, oxygen or other gasses from the air by and into the cryogen causing density variations and corresponding variations in buoyancy.

6. Thermal expansion during warm-up of test mass causing increase in buoyancy in gas or liquid.

C. Seismic/Vibration

1. Local seismic noise effecting one part of a balance preferentially.

2. Subtle seismic or structural vibrations serendipitously synchronized to the expected experimental effect being measured interpreted as signal of real effect. This is especially true for condenser and other sensitive microphones due to high sensitivity over wide frequency response, which are clamped to a laboratory structure.

3. Vibrations from local rotating machines, e.g., roughing and turbomechanical pumps.

D. Diurnal & Gravitational

1. Effect of motion of moon on sensitive balances.

2. Tidal motions of earth's crust altering orientation or periodicity of observations.

3. Cautions regarding use of sealed gravimeters for force detection (placement with respect to experiment, size of internal detection mass, handling, temperature, etc).

4. Avoidance of moving masses in laboratory (e.g., people and equipment) during sensitive gravity experiments.

5. . Calculation of effect of large nearby stationary masses.

E. Vacuum

1. Outgassing of materials in vacuum interacting with movable masses.

2. Outgassing of fastening/joining methods, e.g., gas from blind bold holes interacting with movable masses.

3. Mechanical strains on structural/electrical/measuring components during pump–down.

4. Slow leaks resulting in air stream impacting test mass.

5. Internal "wind" during pump–down or gas back fill.

F. Coriolis/ Earth Rotation, Torques

1. Correction for Coriolis acceleration/earth rotation effects in extremely sensitive moving-mass force–detection systems.

2. For test masses firmly fixed to a balance arm without provision for pivoting or gimbaling, the mass can exert a torque on the arm masquerading as a weight change. Especially true if mass has a magnetic moment (conductor or non-conductor, magnetic or non-magnetic) either induced or permanent, then stray fields can induce a "magnetic" torque in the test mass.

G. Liquid

1. Noise induced in weight/force measuring instruments due to separation of tests mass from liquid (usually cryogen) bath while lifting mass out of bath.

2. Noise induced in weight/force measuring instruments due to evaporation of liquid (usually cryogen) from surface of test mass.

3. Weight artifacts induced in suspended test masses approaching cryogenic temperatures due to condensation of residual water vapor on test mass and suspension not removed by vacuum system. This effect can appear as an increasing weight over time as more water vapor condenses.

4. Artifactual and fluctuating weight changes due to de–wetting of suspension (usually wire or filament) of submerged test mass while surrounding liquid evaporates and level decreases. This effect increases with surface tension, test mass circumference, and decreases with increasing contact angle. Surface roughness also important.

5. Surface tension can exert undesirable forces on a test mass when it passes through the surface of a liquid.

II. Temporal Effects:

A. Signal Duration

1. Mismatch between time scale/time constants of measuring device vs experimental variable.

2. Long-duration signals lost in long-term natural drift of experimental parameters.

B. Test Mass Conditioning

Allowance of sufficient time for sample to reach required temperature (e.g., cooling a superconductor to below transition temperature) between measurements if direct temperature determination is difficult or impossible.

III. Electromagnetic Effects:

A. Magnetic Coupling

1. Influence of time–varying fields on non–magnetic but conducting bodies, inducing local magnetic fields in conducting bodies which may be attracted or repelled from the field or other nearby bodies.

2. Simple magnetic coupling between magnetizable bodies considered unmagnetized before the experiment.

3. Over–reliance on magnetic shielding material which needs special handling and re–annealing after machining/forming/bending etc.

4. Improper reliance on magnetic shielding material for exclusion of DC or quasi–static magnetic fields.

5. Influence of earth's static magnetic field strength, gradient, and dip on magnetic bodies.

6. Stray artificial magnetic fields causing spurious electron beam deflection on oscilloscopes.

7. Sudden release of trapped magnetic fields in superconductors raised above transition temperature affecting & affected by nearby magnetic or conductive structures.

8. Coupling between magnetic moment of superconducting test mass and external magnetic fields including earth's field.

B. Electric/Magnetic Screening

1. Leaking/improperly sealed Faraday Cage/electrostatic screens.

2. Improper reliance on Faraday Cage for complete exclusion of DC or quasi-static electric fields.

3. Frequency dependence of Faraday Cage

4. Inability of screen-type Faraday Cage to screen magnetic fields

C. Electromagnetic Coupling

1. Avoidance of switching transients especially in high–power circuits, especially sudden stopping of current though inductive loads or conductors producing EMP inducing large spurious signals even through shielded coax or aluminum instrument boxes/cases.

2. High frequency RF radiation from nearby transmission lines or conductors interfering with electronic weigh scales.

3. Lack of RF suppression on instrument power lines and instrument lines, e.g., ferrites, shunting caps, proper RF connectors & cables, unless disallowed for frequency response reasons.

4. Avoidance of capacitive coupling between signal cables and grounds/ground leads carrying transient/fault currents.

5. When a source is incorrectly matched to a load, a greatly increased level of EMI across a broad frequency range may be generated as the reflected power interferes with the correct operation of the source (an amplifier usually). This in turn may cause spurious measurements to occur, and this is particularly troublesome when using an electronic balance.

6. Casimir force between test mass and measuring system at nanoscale dimensions.

D. Grounding/Earthing

1. Avoidance of contact potentials developing across multiple ground connections. In some cases contact potentials must be compensated by a deliberately applied counter potential.

2. Strive for single–point RF ground system for all instruments.

3. Correction of ground loops and ground faults both internal to the experiment and between experiment and measuring system.

4. Understanding the difference between independent earth ground (e.g., copper stake in virgin earth) vs mains "ground" vs mains neutral, and potentials between these.

5. Poor/loose ground connections: preventing complete charge draining, allow transient voltage artifacts on recording & display devices, allow small signals to be amplified by amplifiers along with the signal of interest, etc.

6. Use of large cross-section circular wire or flat ribbon strip from experiment and/or instrumentation to earth especially for pulsed high–power experiments.

IV. Electrostatic and Related Effects:

A. Gradient

Gradient of electrostatic field caused induced motion in nearby free bodies.

B. Charge Pooling & Induced Charges

1. Accumulation of pools of surface charges on invisible insulating patches on conductors. Especially problematic for metal enclosures/surfaces which have unavoidable insulating metal oxide layer formed on surface, eg. aluminum.

2. Accumulation of surface charges on water patches on inner surfaces of vacuum chambers and components.

3. Accumulation of charge on insulating or non-conductive surfaces, e.g.,wire insulation, after exposure to electrostatic and sometimes time-varying electric fields.

4. Reaction against image charges created on conductor.

C. **Ion & Molecular**

1. "Ion wind" due to ionized surrounding gas causing artificial force on conductors especially in high-voltage DC or AC experiments.

2. High-voltage ablation/sputtering of molecules or ions from conductors or insulators.

D. **Charge Leakage**

1. Unaccounted–for corona or other uncontrolled charge leakage, usually in bursts ("Tricel Pulses") in high-voltage experiments, which can create time–varying charge on nearby conductors. Especially problematic at sharp corners.

2. Ions from leakage current interacting with gas molecules and imparting thrust to the leakage ion source body.

3. High voltage creation of weak conduction paths between device under test and ground. Depends on humidity, vacuum.

V. **Instrumentation Issues**

1. Measurement outside specifications of instruments, including sensing and measuring instruments; signal processors, amplifiers, and conditioners; and recording, display, and acquisition devices.

2. Lock–in amplifier response to high-amplitude transients riding on input lines, causing artifacts even when not phase locked to the reference signal.

3. Voltage sags/surges resulting in poor mains power quality, e.g., startup of nearby large rotating equipment.

4. Ensuring correct vertical/plumb orientation of torsion balances, especially while on pneumatic anti–vibration tables.

5. Operation of bearings & pivots outside specifications.

6. Operation of bearings & gears in vacuum using proper vacuum grease.

VI. **Signal Analysis**

1. Averaging: to tease out buried signals and suppress noise

2. Statistical Analysis: use of χ^2, calc. of correlation coefficients, sigmas, etc.

3. Noise: is noise floor burying signals of interest?

4. Error Analysis: how confident that signal is inside measuring instrument range and is real – requires full specs. of instrumentation, error propagation.

5. Exploiting Adjustable Parameters

    - Adjusting phase of various parameters to detect artifacts.
    - Suppression of common–mode noise.
    - Alternate mechanical orientation of experiment with respect to possible local forces or gravity.

VII. Use of dummy test mass/controls

1. Replacement of test mass by known null-effect mass.

2. Use of null-effect dummy mass of identical thermal characteristics to test mass, especially in superconductor/cryogen experiments.

3. Reversing sense of one experimental variable to determine if observed effect goes away.

4. Shorting one component or use of dummy electrical component with same electrical characteristics.

5. When using superconductors as test mass, check for correlations with test mass being in Meissner state.

6. Investigate all properties of test mass before and after experiment, including volume properties (phases, crystal or amorphous structure, chemical composition, absorbed species, density, thermal diffusivity) and surface properties (morphology, interfacial energy, wettability, depositions/adsorption, corrosion and erosion).

Finally, since this is very important to people at this workshop, here are some issues relevant to Roger Shawyer's testing of the recently-announced "EM Drive", taken from the list above .

- Self-contained power supply, or wires to lab frame?
- Absolutely vertical thrust bearing shaft?
- Rigidity of support frame?
- Off-center center-of-mass at any time during the test? changes in COM?

- Evidence of ratcheting? ("Dean Drive" effect)
- Serious thermal issues including waste heat flow, differential thermal expansion causing variations in lever arm

And some control experiment suggestions:

- Replacing thruster with cylindrical cavity and/or back-to-back tapered cavities
- De-Q the cavity tapered walls
- Re-arrange placement of radiator & piping vs cavity
- Re-arrangement of wires

References

[1] Peter Graneau PhD., "Extracting Intermolecular Bond Energy from water", http://www.infinite-energy.com/images/pdfs/GraneauIE13-14.pdf

[2] G. Hathaway & B. Cleveland, "Gravity modification experiment using a rotating superconducting disk and radio frequency fields", Physica C Superconductivity **385**(4):488-500 April (2003)

[3] Dr. Evgeny Podkletnov on the Impulse Gravity-Generator, on http://energia.moebius.com.br/oo/Energia/ENERG4L-antigravity/EugenePodkletnov-RussianScientist/PodkletnovInterview(EN).pdf see also Evgeny Podkletnov and Giovanni Modanese, *Gravity-Superconductors Interactions: Theory and Experiment* Bentham science books/9781608053995.

[4] Rudolph G. Zinsser, "Mechanical Energy from Anistropic Gravitational Fields", http://www.rexresearch.com/zinsser/zinsser.htm US Patent 4,085,384.

[5] G. Hathaway, *Mindbending - The Hutchison Files: 1981 to 1995*, Paperback –July 25, (2014).

[6] J. F. Woodward, "*Making Starships and stargates*", (Springer press Dec 2012).

[7] G. Hathaway, "Sub-micro-Newton resolution thrust balance", Rev. of Sci. Instr. **86**, 105116 (2015).

Estes Park Quick Study IV

Maxwellian Gravity

Reference: *Spacetime and Geometry*, by Sean Carroll, 2004, section 7.2

General Relativity (GR) is our theory of gravity, space, and time. The theory has been confirmed experimentally in many different ways. Yet the equations are enormously complex. Specifically, they are non-linear. Unlike the Maxwell equations, there are no systematic techniques to obtain solutions to the Einstein field equations. All solutions – such as the Robertson-Walker metric, the Schwarzschild metric, the Kerr metric, etc – are guessed and then verified. This means there may well be complexity in the field equations that we have not yet discovered.

Therefore, linear solutions have been sought to the equations since Einstein. To linearize GR, one expresses the full spacetime metric $g_{\mu\nu}$ into a perturbation $h_{\mu\nu}$ about the flat space metric $\eta_{\mu\nu}$, such that $g_{\mu\nu} = \eta_{\mu\nu} + h_{\mu\nu}$, $h_{\mu\nu} \ll 1$. This expression is plugged into the equations of GR and terms are only kept to linear order in $h_{\mu\nu}$.

Carroll suggests decomposing the components of $h_{\mu\nu}$ according to their spatial transformation properties. Then $h_{00} \equiv \phi$ transforms as a scalar, and $h_{0i} \equiv \mathbf{A}$ transforms as a vector. The remaining components are the 3x3 matrix h_{ij}. This decomposition is entirely analogous to decomposing the electromagnetic field strength tensor into electric and magnetic field vectors.

For a particle of energy E and velocity $\mathbf{v}$ moving in this gravitational field, its energy equation is (Carroll 7.23)

$$\frac{dE}{dt} = -E\left[\frac{\partial \phi}{\partial t} + 2\mathbf{v} \cdot \nabla\phi - \mathcal{O}(v^2)\right]$$

Now define gravito-electric field $\mathbf{G}$ and gravito-magnetic field $\mathbf{H}$:

$$\mathbf{G} \equiv -\nabla\phi - \frac{\partial \mathbf{A}}{\partial t}$$

$$\mathbf{H} \equiv \nabla \times \mathbf{A}$$

Then the particle spatial momentum $\mathbf{p}$ is described by (Carroll 7.26):

$$\frac{d\mathbf{p}}{dt} = E\left[\mathbf{G} + \mathbf{v} \times \mathbf{H} - \mathcal{O}(\partial h_{ij}/\partial t, v^2)\right]$$

These are clearly very similar to the Lorentz force law of electromagnetism, and one can infer that GR includes magnetic-type gravitational forces. Unlike the Maxwell equations, however, we also have additional, non-linear terms.

Interestingly, Carroll goes on to show how the only components with true degrees of freedom are the spatial components h_{ij}. The suggestive gravito-electric and gravito-magnetic fields are fixed by the stress energy tensor and the h_{ij}. Carroll goes on to choose a particular gauge for the $h_{\mu\nu}$, and to show the h_{ij} is the piece that contains gravitational radiation.

L.L. Williams
July 2016

Inertia from Gravity: Insights of Sciama

Reference: *On the Origin of Inertia*, D.W. Sciama, MNRAS, 113, 1953

The fuel problem (see EP Quick Study I) can also be seen as a problem of inertia. If it weren't for inertia, we would not need any fuel to push a spaceship. The resistance to acceleration comes from inertia.

We may expect there to be some link between gravity and inertia, since they have the same mass parameter. While Einstein built General Relativity on the equivalence of gravitation and inertia, the origin of inertia is still debated. Insight into the origin of inertia came from a simple argument by Sciama. This insight has formed a cornerstone of Woodward's Mach Effect theory.

Sciama quantifies inertia by assuming gravity is a vector field, and using the well-known mathematics of electrodynamics to quickly derive its inertial implications. We know, and Sciama knew, that gravity is a tensor field, but it is a reasonable approximation because Sciama's result only requires that gravity be at least as complex as a vector field. Einsteins equations of gravity can be linearized, and they do indeed show a Maxwellian character (see EP Quick Study IV). There are of course the gravito-electric effects one expects for Newtonian gravity, but other gravito-magnetic effects that Newton would not have recognized. These latter effects are more conventionally known as frame dragging.

So Sciama starts with a Maxwellian gravitational 4-vector potential $(\phi, \mathbf{A})$. The scalar potential has the usual mathematical form, but in terms of mass density ρ instead of charge density, and in terms of the gravitational constant G instead of the dielectric constant:

$$\phi = -G \int \frac{\rho}{r} dV$$

The integral is taken over the whole universe, but we know the universe is undergoing a Hubble expansion of speed $v_H = Hr$, where H is the Hubble parameter. The current value of $H_0 \simeq 70$ km/s per megaparsec. It is presumed no part of the universe receding faster than the speed of light can influence the local scalar field of gravity, so the integral is cut off at a distance $r_H = c/H$. The value of H changes over time, and the horizon distance is the farthest object to have emitted light just now reaching us. The horizon distance depends on the energy content of the universe, but it scales with the instantaneous horizon distance $r_H = c/H \sim c/H_0$. Therefore we can approximate the scalar potential of the universe:

$$\phi \simeq -G \int_0^{r_H} \frac{\rho}{r} 4\pi r^2 dr = -2\pi G \rho_U \left(\frac{c}{H_0}\right)^2$$

where we have approximated the mass density of the universe as a constant ρ_U. This equation already contains a compelling feature. If taken at infinity, the integral would diverge quadratically. This means the inertia here is dominated by matter in the distant universe.

We construct the usual spatial vector potential in terms of a mass current **J**:

$$\mathbf{A} = -\frac{G}{c^2} \int \frac{\mathbf{J}}{r} dV$$

At this point, Sciama invokes Mach's principle to presume that there is a rest frame for the universe, and in that frame the current must be zero. But for an object in motion with respect to the rest frame of the universe, the universe appears to be in motion and the object at rest. Therefore an apparent current arises in the universe from the motion of the object. In this case, the current of the universe due to the motion **v** of the object is $\mathbf{J} = \rho \mathbf{v}$. The gravitational vector potential of the universe is then:

$$\mathbf{A}_U = -\frac{G}{c^2} \int_0^{r_H} \frac{\rho \mathbf{v}}{r} 4\pi r^2 dr = \phi_U \frac{\mathbf{v}}{c^2}$$

Construct the usual Maxwellian gravito-electric force from the potentials:

$$\mathbf{f} = -\nabla \phi_U - \frac{\partial \mathbf{A}_U}{\partial t} = 2\pi G \frac{\rho_U}{H_0^2} \frac{\partial \mathbf{v}}{\partial t}$$

Now the identification of inertia is complete, if the coefficient leading the derivative of the velocity is unity. Remarkably – and unknown in Sciama's time – cosmology tells us it is.

The Friedmann equation is the workhorse of modern cosmology. It relates the expansion of the universe $H(t)$ to curvature κ and energy density $\epsilon(t)$:

$$H^2(t) = \frac{8\pi G}{3c^2} \epsilon(t) - \frac{\kappa c^2}{R_0^2 a^2(t)}$$

For a flat universe, $\kappa = 0$ and the energy density of the universe is the critical density:

$$H^2(t) = \frac{8\pi G}{3c^2} \epsilon_c(t)$$

In fact, modern cosmology tells us that the universe is flat and the equation above relates the Hubble parameter to the energy density of the universe. If we set $\epsilon_c = \rho_U c^2$ with the understanding that ρ_U includes all the gravitating mass-energy in the universe, we find

$$H_0^2 = \frac{8\pi G}{3} \rho_U$$

and therefore

$$\mathbf{f} = \frac{3}{4} \frac{\partial \mathbf{v}}{\partial t}$$

The coefficient is close enough to unity that we would have to revisit our approximation to GR, our integration limit, and our detailed model of the expansion of the universe, to definitively rule out gravity as the origin of inertia. Modern cosmology and the flat universe appears to reinforce the conclusion that inertia can be accounted for by the gravitational influence of the universe.

A closely related result of the flat universe is that the gravitational potential energy of every particle in the universe exactly equals its rest mass energy. That is, $\phi = c^2$. Therefore, the speed of light is set by the gravitational potential of the universe.

L.L. Williams
July 2016

Estes Park Quick Study VI

Woodward's Mach Effect Equation

This is a summary of Woodward's own derivation of the Mach equation, as given in his book *Making Starships and Stargates*. It is compressed here to facilitate comparison and analysis at Estes Park.

Woodward starts with the observation from Sciama, 1953, that gravitational forces from the rest of the universe can account for inertia (cf. EP Quick Study V). The inertial force is per unit mass, like a gravitational field; call it **f**.

A relativistic description of force involves the proper time derivative of the 4-momentum:

$$F^\mu = \frac{dp^\mu}{d\tau} = \left(\frac{dp^0}{d\tau}, \frac{d\mathbf{p}}{d\tau}\right) \tag{1}$$

This is a 4-vector expression. Greek indices range over the 4 coordinates of spacetime. The time component of the 4-vector is noted with a 0 index, and the 3 spatial components are written as a vector in boldface. The time component of the 4-momentum is the energy. Therefore a 4-vector that corresponds to the inertial force **f** per unit mass will have a time-component equal to the work done on the object by the force, per unit mass:

$$\frac{F^\mu}{m} = f^\mu = \left(\frac{1}{m}\frac{dp^0}{d\tau}, \mathbf{f}\right) \tag{2}$$

Woodward elects to investigate the 4-divergence of the relativistic inertial force field:

$$\nabla_\mu f^\mu = \frac{1}{c}\frac{\partial}{\partial t}\left(\frac{1}{m}\frac{dp^0}{d\tau}\right) - \nabla \cdot \mathbf{f} \tag{3}$$

Woodward posits that the divergence (3) of the inertial force must correspond to a field equation. He sets the right hand side equal to some unspecified source $4\pi Q$, as might be expected for such an equation. He converts the particle energy into an energy density E, to conform to the expected units in a field equation. And he considers non-relativistic speeds, so that the proper time derivative becomes a simple time derivative, and the energy density is simply the mass density time the speed of light squared, c^2:

$$\frac{1}{c^2}\frac{\partial}{\partial t}\left(\frac{1}{\rho}\frac{\partial E}{\partial t}\right) - \nabla \cdot \mathbf{f} = 4\pi Q \tag{4}$$

The first term on the left side expands into 2 terms, differing by a minus sign due to the derivative of the inverse mass density. Now

Woodward makes 3 simultaneous ansatzes, one for each term of equation (4):

1. The source term is the usual source term of Newtons law of gravity: $Q \to G\rho$, the product of mass density and Newton's gravitational constant
2. The inertial reaction force $\mathbf{f} \to -\nabla \phi(\mathbf{x}, t)$, the gradient of the gravitational potential in Newton's law of gravity
3. The energy density $E \to \rho(\mathbf{x}, t)\phi(\mathbf{x}, t)$. This is the Mach hypothesis: that the energy of a body depends on the gravitational potential. Equivalently, the speed of light squared is the same as the value of the scalar potential, $c^2 \to \phi(\mathbf{x}, t)$.

With these assumptions and a prescription for where to make the substitutions, Woodward calculates his Mach effect equation, which is a modification of Newtons field equation:

$$\nabla^2 \phi - \frac{1}{c^2} \frac{\partial^2 \phi}{\partial t^2} = 4\pi G \rho + \frac{\phi}{\rho c^2} \frac{\partial^2 \rho}{\partial t^2} - \left(\frac{\phi}{\rho c^2}\right)^2 \left(\frac{\partial \rho}{\partial t}\right)^2 - \frac{1}{c^4}\left(\frac{\partial \phi}{\partial t}\right)^2 \quad (5)$$

The basic feature of Woodward's equation (5) is time-derivatives of the field acting as a source, and these sources are scale free: they dont involve the gravitational constant. So they are relatively larger than the conventional source term, the first term on the right side. They also have the negative-definite terms that Woodward would suggest for creating warp drives or wormholes.

As Woodward notes, equation (5) does not tell us anything about how the mass time derivatives come about. Naively it would appear to say a change in internal energy of any sort could bring about this effect. But Woodward hearkens back to the original assumption, we could call it Woodwards zeroth ansatz, that inertial reaction forces arise from the rest of the universe resisting the acceleration of a body. If the body is not accelerated, no inertial reaction force is raised.

Woodwards 4th ansatz is that the change in internal energy scales with the acceleration that raises the inertial reaction force. So he then equates the inertial reaction force with the bulk acceleration of the object:

$$\Delta E = m\mathbf{f} \cdot \Delta \mathbf{s} = m\mathbf{a} \cdot \Delta \mathbf{s} \quad (6)$$

The quantity $\Delta \mathbf{s}$ is a parameterization of the work done by the inertial reaction force, with units of length. It is understood that this does *not* correspond to the bulk displacement of the object, but rather, to some internal dissipation. This gives

$$\frac{\partial E}{\partial t} = m\mathbf{a} \cdot \frac{\partial \mathbf{s}}{\partial t} \quad (7)$$

and

$$\frac{\partial^2 E}{\partial t^2} = m\mathbf{a} \cdot \frac{\partial^2 \mathbf{s}}{\partial t^2} + \frac{\partial \mathbf{s}}{\partial t} \cdot \left(m\frac{\partial \mathbf{a}}{\partial t} + \mathbf{a}\frac{\partial m}{\partial t}\right) \quad (8)$$

Finally, Woodward hypothesizes that the parameterization of dissipation $\partial \mathbf{s}/\partial t$ must scale with the bulk velocity $\mathbf{v}$, so that $\partial \mathbf{s}/\partial t = \eta \mathbf{v}$. Then, since $\mathbf{a} = \partial \mathbf{v}/\partial t$, Woodward finds (3.7) of his book:

$$\frac{\partial^2 E}{\partial t^2} = \eta m a^2 + \eta \mathbf{v} \cdot \left(m \frac{\partial \mathbf{a}}{\partial t} + \mathbf{a} \frac{\partial m}{\partial t} \right) \qquad (9)$$

At successive points in Woodwards development, he drops the second term as much smaller than the first term in the equation above, and also drops the term quadratic in $\partial m/\partial t$ in (5). Adopting these approximations now, and putting all this together, we obtain the approximate equation for the mass fluctuation:

$$\delta m = V \delta \rho = \frac{\eta}{4\pi G} \frac{V a^2}{c^2} = \frac{\eta}{4\pi G} \frac{m a^2}{\rho c^2} \qquad (10)$$

which is equation (5.9) from Woodwards book, and where the mass fluctuation is defined in terms of the standard Newtonian expression:

$$\nabla^2 \phi = 4\pi G (\rho + \delta \rho) \qquad (11)$$

This is apparently the mass fluctuation that is produced from accelerating objects, and which Woodward would hope to engineer into various designs that time the mass fluctuations with internal constituent motions of the thruster to produce a net impulse.

Note that the gravitational constant enters inversely, which presumably results in very large effects.

L.L. Williams
July 2016

| ✓ Fuel Problem | ✓ Theory | ✓ Existing Physics |
| ✓ Time-Distance Problem | ✓ Experiment | New Physics |

Mach Effect Thrusters

- Jim Woodward and Heidi Fearn will lead a discussion about theory and experimental efforts regarding a Mach Effect thruster
- Heidi Fearn and José Rodal will lead a discussion regarding Mach Effects interpreted within the Hoyle-Narlikar theory. Heidi will give an overview, and José will discuss a particular application.

Issue Summary:

This is the most well-developed topic of the conference, because there is substantial development in both theory and experiment. Interestingly, it does not invoke any new physics, although experimental validation would be a great discovery. The Mach Effect thrusters exploit an understanding of inertia that lies within the framework of our current understanding of gravity; see the Estes Park Quick Study on Sciama's insight into inertia.

The theory and first experiments were developed by Jim Woodward at Cal State Fullerton. He has been collaborating in recent years with Heidi Fearn at Fullerton, and other groups have performed Mach thruster experiments. Since many of you know Jim Woodward personally, you may have access to his book [1], so that is recommended as an overview of Woodward's theory and experiment. A key paper is here [2]. A digest of Woodward's mathematical theory lifted from Woodward's book is included as Estes Park Quick Study VI, to facilitate analysis of Woodward's theory. One finds that while the derivation is reasonable, an alternative derivation suggests itself within standard techniques of linearized GR.

Woodward maintains that the Mach Effect addresses both the fuel problem and the time-distance problem. The latter is because there are negative-definite source terms in the field equations that are suggestive of the negative energies posited by Kip Thorne's worm hole analysis.

Jim and Heidi will tag-team the discussion of theory and experiment. One objective for the theory part of the session is whether the Mach Effect theory can be more firmly embedded in the techniques of GR. Another objective is to gain a clear understanding of the experimental set up. The implications for a Mach interpretation of other experiments will be discussed.

Heidi Fearn and José Rodal will tag-team discussions on Mach effect applications within the theory of Hoyle and Narlikar from 1964. See [3] and [4]. The Hoyle-Narlikar theory is closely tied to the old Steady-State cosmology of the 1960s. Now, in the age of precision cosmology, the Lambda-Cold-Dark-Matter model [5] is parameterized to 3 or 4 significant figures, and Steady-State cosmologies are no longer considered. But Heidi and José will share some compelling experimental implications of the theory, which they feel addresses some causal aspects of the coupling of inertia to the distant universe. They are particularly intrigued by the concept of Wheeler-Feynman absorber theory [6] and how it applies to inertia.

1. Making Starships and Stargates, ISBN 978-1-4614-5622-3, Springer: New York, 2013.
2. Flux capacitors and the origin of inertia, Found. Phys., v34, p1475, October 2004.
3. F. Hoyle & J. Narlikar, PRSL-A, v282, p191, November 1964
4. F. Hoyle & J. Narlikar, PRSL-A, v277, p1, January 1964
5. J. Frieman, M. Turner, & M. Huterer, Ann. Rev. Astron. Astrophy., v46, p385, 2008.
6. J. Wheeler & R. Feynman, Rev. Mod. Phys., v17, p157, 1945.

Gravitational Absorber Theory & The Mach Effect

H. Fearn
Physics Dept. CSU Fullerton,
800 N. State College Blvd.
Fullerton CA 92834, USA

The origins of mass can be described in terms of Mach's principle, which states that the mass of a body is determined by its interaction with the rest of the mass-energy in the universe. However, if a body undergoes a sudden acceleration, you may ask, "How can the universe respond immediately in a way to conserve momentum?" In order to explain this, we introduce the concept of advanced waves, which have been used successfully in both classical and quantum physics for the last 70+ years.

1. Introduction

The work of Hoyle and Narlikar (HN) is a "masterpiece" in general relativity (GR) theory because it is fully covariant and incorporates fully the idea of Mach's principle. It is what Einstein dearly wanted to do, but didn't think he quite managed with standard general relativity. But perhaps he did? This paper shows that the results of HN, or what I would prefer to call *gravitational absorber theory* (GAT) can be obtained from Einstein's GR with the addition of a mass fluctuation in time. In Section 2, I show that adding $m(t)$ is all that is needed. I have renamed the theory to emphasise that I am not interested in the static universe model; I do not include the HN mass creation (C)-field. I am only interested in the Machian development of the theory through the use of retarded and advanced waves.

The local field around a mass particle can still be thought of as the overlapping of the many retarded and advaced waves, which themselves carry energy and momentum. This field will have a potential anywhere in space-time and constitutes the background vacuum. The mass particle transfers energy and momentum with the "field" here and now, which is basically th vacuum. However, when the particle accelerates the universe as a whole reacts to the acceleration, causing changes in the local field, which can be transmitted to the particle conserving mometum on a universal scale.

Einstein began his work on general relativity by seeking a concordance with Mach's principle. That is, to explain inertia of a test mass in terms of other masses in the universe. Sciama, Nordtvedt, and others have shown that masses in motion exert non-radial gravitational forces on nearby masses (frame dragging). In particular, Sciama showed that just this frame-dragging effect from the rest of the universe can account for inertia. Woodward has exploited the result of Sciama to design a propellantless propulsion device that depends on such forces.

In spite of its prediction of frame-dragging, and apparent ability to account for inertia, some researchers feel that general ralativity does not provide a fully self-consistent Machian picture. While Sciama and Nordtvedt can calculate the inertial force on an accelerated object due to the rest of the mass in the universe, we feel that general relativity does not account for the effect of the accelerated mass back on the rest of the universe. To properly account for the effect of the accelerated mass back on the rest of the universe, we employ the concept of advanced waves, made famous by Wheeler and Feynman. Hoyle and Narlikar developed a theory of general relativity that incorporated advanced waves. While Hoyle and Narlikar are well-known for their steady-state cosmology work, and they use HN theory in that work, we feel their theory stands as a fine extension to general relativity, ignoring the parts regarding mass creation.

The beauty of gravitational absorber theory (GAT) is that it allows one to think of a mass, here and now, being influenced by the rest of the matter in the universe via gravitational signals travelling at speed c and does not rely on some (old fashioned Newtonian) notion of "action-at-a-distance" or faster than light propagation of signals. Real gravitational signals travelling at speed c carry information from every part of the universe to a mass here and now. The only trick is, to have mass react instantaneously, you must invoke the advanced wave solution to the relativistic wave equation. This advanced wave travels backward in time from the distant reaches of the universe, to convey momentum to the here and now, allowing back reaction to appear instantaneous.

The reason that HN theory did not catch on in the 1960's is twofold.

1. Hoyle was looking for a static universe cosmology theory. He introduced the "C" field as a creation field to keep the mass density constant as the universe expanded. This C field can be removed without loss of the underlying theory.

2. Hawking raised an objection to the HN theory in 1965 which basically put the last nail on the coffin. He suggested that by integrating out into the distant future, the advanced wave integrals would diverge. That is correct. However, since the universe is not only expanding but accelerating in that expansion, there is a cosmic horizon beyond which you cannot integrate. That cutoff prevents the advanced wave integrals from diverging and therefore re-establishes

the HN theory as a good working theory.

3. Now is the time to look at gravitational absorber theory in a new light. Forget the static cosmology and move forward.

Standard GR has the problem that masses are treated as static. That is in general not the case. The background gravitational potential can be nonzero even in a flat spacetime. The GAT allows for a dynamic communication of signals from every part of the universe to the here and now to conserve momentum. *Furthermore, and this is conjecture, the superposition of retarded and advanced waves throughout the universe could be a mechanism to understand dark energy and matter. For example, dark matter might just be the manifestation of the gravitational potential at a location in space where the retarded and advanced waves do not perfectly overlap. For example at the location of an accelerating mass. As an electromagnetic analogy, consider photons appearing near an accelerating mirror in the dynamic Casimir effect or equivalently Unruh radiation.*

There is sufficient reason to reconsider the gravitational absorber theory of Hoyle and Narlikar. In section 2, we allow for a mass fluctuation in the Einstein equation of motion (geodesic) and obtain the HN equation of motion, which is a new result. In section 3, we give a very brief history of the HN paper sequence and rewrite their notation to assist the reader. In section 4, we compare the Einstein action and field equation with the HN field equation. In section 5, we show that the mass fluctuation frmula calculated by Woodward from the precepts of general relativity can also be obtained from HN theory. This is the main result of the paper.

Advanced waves were introduced by Dirac in 1938 to describe radiation reaction. His radiation reaction force equation is still in use today and can be found in most standard electrodynamics text books. The advanced wave concept was given a physical interpretation by Wheeler and Feynman in 1945 [1]. The idea has since been used successfully in quantum mechanics by John Cramer and later in the theory of gravitation by Hogarth 1962 [2] and Hoyle and Narlikar 1964 [3,4] whose work we will summarize for convenience below.

1.1 Electron Radiation Reaction in Electrodynamics

Dirac [5] first introduced the idea of advanced waves in electromagnetism in order to derive the radiation reaction of an accelerating electron.

The idea is as follows, consider a single electron undergoing acceleration. The field surrounding the electron can be thought of in two parts, the outgoing and incoming. The actual field surrounding the electron is the usual retarded Lienard-Wiechert potentials and any incident field on the electron.

$$F^{\mu\nu}_{act} = F^{\mu\nu}_{ret} + F^{\mu\nu}_{in} \tag{1}$$

Furthermore, the Maxwell 4-potential wave equation allows for advanced solutions, which are the same form as retarded, only they go backward in time. The advanced solutions also satisfy the wave equation in Lorentz gauge (below, with $c = 1$):

$$\Box A_\mu = 4\pi j_\mu \qquad (2)$$

$$\frac{\partial A_\mu}{\partial x_\mu} = 0 \qquad (3)$$

We could equally well describe the actual field surrounding the electron by

$$F^{\mu\nu}_{\text{act}} = F^{\mu\nu}_{\text{adv}} + F^{\mu\nu}_{\text{out}} \qquad (4)$$

where the $F^{\mu\nu}_{\text{out}}$ is the total field leaving the electron. The difference between the outgoing waves and the incoming waves is the radiation produced by the electron due to its acceleration.

$$F^{\mu\nu}_{\text{rad}} = F^{\mu\nu}_{\text{out}} - F^{\mu\nu}_{\text{in}} = F^{\mu\nu}_{\text{ret}} - F^{\mu\nu}_{\text{adv}} \qquad (5)$$

In the appendix of Dirac's paper, it is shown that this equation gives exactly the well known relativistic result for radiation reaction which can be found in standard text books on electromagnetism, for example Jackson [6].

1.2 Wheeler & Feynman: Absorber Theory

Wheeler and Feynman [1] accept Dirac's result but wish to give a physical explanation as to where the advanced electromagnetic field comes from. They resort to a suggestion made by Tetrode [7] and later by Lewis [8] which was to abandon the concept of electromagnetic radiation as a self interaction and instead interpret it as a consequence of an interaction between the source accelerating charge and a distant absorber. The absorber idea has the four following basic assumptions, which we quote directly from Wheeler-Feynman [1],

> (1) An accelerated point charge in otherwise charge-free space does not radiate electromagnetic energy.
> (2) The fields which act on a given particle arise only from other particles.
> (3) These fields are represented by 1/2 the retarded plus 1/2 the advanced Lienard-Wiechert solutions of Maxwell's equations. This force is symmetric with respect to past and future.
> (4) Sufficiently many particles are present to absorb completely the radiation given off by the source.

Now, Wheeler-Feynman considered an accelerated charge located within the absorbing medium. A *disturbance* travels outward from the source. The absorber particles react to this disturbance and themselves generate a field half advanced and half retarded waves. The sum of the advanced and retarded effects of all the charged particles of the absorber, evaluated near the source charge, give an electromagnetic field with the following properties [1]:

(1) It is independent of the properties of the absorbing medium.
(2) It is completely determined by the motion of the source.
(3) It exerts on the source a force which is finite, is simultaneous with the moment of acceleration, and is just sufficient in magnitude and direction to take away from the source the energy which later shows up in the surrounding particles.
(4) It is equal in magnitude to 1/2 the retarded field minus 1/2 the advanced field generated by the accelerated charge. In other words, the absorber is the physical origin of Dirac's radiation field
(5) This field combines with the 1/2 retarded, 1/2 advanced field of the source to give for the total disturbance the full retarded field which accords with experience.

The Wheeler-Feynman paper presents four derivations of the relativistic radiation reaction of an accelerated charge, each successive derivation increasing in generality. The first three derivations proceed by adding up all the electromagnetic fields due to the absorber particles. The fourth is the most general derivation, which only assumes that the medium is a complete absorber and so outside the medium the sum of all the retarded and advanced waves is zero. Each yields the well-known relativistic radiation reaction as given in text books [6].

So far, we have shown that the advanced wave idea has been used successfully in classical physics. Now we proceed to show that it can also be advantageously used within quantum mechanics. The transactional interpretation of quantum mechanics was written by John Cramer [9,10] in the 1980's. It is a way to view quantum mechanics which is very intuitive and easily accounts for all the well known paradoxes, EPR, which-way detection and quantum eraser experiments. We refer the reader to his paper, which is a very interesting read. All the usual quantum results hold, and it is simply an alternative point of view from the Copenhagen interpretation and collapsing-wave-function way of thinking.

Hoyle-Narlikar (HN) theory is a kind of absorber theory (with advanced waves) for gravitation rather than electrodynamics. HN theory agrees with Einstein's theory of gravitation in the limit of a smooth fluid mass density distribution in the rest frame of the fluid. All of the tests of Einstein's gravitation still apply to HN theory.

2. Derivation of the Einstein Geodesic Equation

One method used to derive the geodesic equation, also known as the equation of motion of a particle, is extremizing (minimizing) the line element. This will give us the shortest distance between two points. Taking the general line-element,

$$ds^2 = g_{\mu\nu}dx^\mu dx^\nu \tag{6}$$

varying both sides (a similar derivation can be found in Dirac [11] p16), we get

$$\begin{aligned}
2ds\delta(ds) &= dx^\mu dx^\nu \delta(g_{\mu\nu}) + g_{\mu\nu}dx^\mu \delta(dx^\nu) + g_{\mu\nu}dx^\nu \delta(dx^\mu) \\
&= dx^\mu dx^\nu g_{\mu\nu,\lambda}(\delta x^\lambda) + 2g_{\mu\lambda}dx^\mu \delta(dx^\lambda) \\
\delta(dx^\lambda) &= d(\delta x^\lambda) \\
dx^\mu &= \left(\frac{dx^\mu}{ds}\right)ds = v^\mu ds
\end{aligned} \tag{7}$$

In order to extremize the action $\int \delta(mds)$, treat mass as a constant. Then we consider the following,

$$\begin{aligned}
\int \delta(ds) &= \int \left[\frac{1}{2}\frac{dx^\mu}{ds}\frac{dx^\nu}{ds}g_{\mu\nu,\lambda}\delta x^\lambda + g_{\mu\lambda}\frac{dx^\mu}{ds}\frac{d}{ds}(\delta x^\lambda)\right]ds \\
&= \int \left[\frac{1}{2}g_{\mu\nu,\lambda}v^\mu v^\nu (\delta x^\lambda) + g_{\mu\nu}v^\mu \frac{d}{ds}(\delta x^\lambda)\right]ds
\end{aligned} \tag{8}$$

Integrating the second term by parts we find

$$\int \delta(ds) = \int \left[\frac{1}{2}g_{\mu\nu,\lambda}v^\mu v^\nu - \frac{d}{ds}(g_{\mu\lambda}v^\mu)\right](\delta x^\lambda)ds = 0 \tag{9}$$

For this to be true for any variation δx^λ we find that the term inside the square bracket must be zero hence,

$$\frac{d}{ds}(g_{\mu\lambda}v^\mu) - \frac{1}{2}g_{\mu\nu,\lambda}v^\mu v^\nu = 0 \tag{10}$$

Furthermore

$$\begin{aligned}
\frac{d}{ds}(g_{\mu\lambda}v^\mu) &= g_{\mu\lambda}\frac{dv^\mu}{ds} + g_{\mu\lambda,\nu}v^\mu v^\nu \\
&= g_{\mu\lambda}\frac{dv^\mu}{ds} + \frac{1}{2}(g_{\mu\lambda,\nu} + g_{\lambda\nu,\mu})v^\mu v^\nu
\end{aligned} \tag{11}$$

By substitution of Eq. (11) into Eq. (10) we find

$$\begin{aligned}
g_{\mu\lambda}\frac{dv^\mu}{ds} + \frac{1}{2}(g_{\lambda\mu,\nu} + g_{\lambda\nu,\mu} - g_{\mu\nu,\lambda})v^\mu v^\nu &= 0 \\
\frac{1}{2}(g_{\lambda\mu,\nu} + g_{\lambda\nu,\mu} - g_{\mu\nu,\lambda}) &= \Gamma_{\lambda\mu,\nu} \\
\frac{dv^\sigma}{ds} + \Gamma^\sigma_{\mu\nu}v^\mu v^\nu &= 0
\end{aligned} \tag{12}$$

where the last equation, which is the usual geodesic equation, follows when you multiply throughout by $g^{\lambda\sigma}$. We did not start with a true particle Lagrangian, only a line element. The Lagrangian includes a mass of the particle. Note that we have left the mass entirely out of the variation since at present it is treated as a constant. Let us compare this with what happens when you vary the rest mass of the particle.

2.1 Allow the Mass to change in Equation of Motion derivation

To see how similar the HN equation of motion is to the Einstein Geodesic, simply repeat the above calculation but allow the mass to change. The result is the Equation of Motion for the HN-theory. It is a little unclear as to why the usual Einstein geodesic does not contain the same mass variation. Is it because the mass is held constant in the Einstein case? If so **why** is the mass held constant?

Starting from the line element,

$$ds^2 = g_{\mu\nu} dx^\mu dx^\nu$$

$$2ds\delta(ds) = \delta g_{\mu\nu} dx^\mu dx^\nu + 2g_{\mu\nu} dx^\mu \delta(dx^\nu)$$

$$\delta(ds) = \left[\frac{1}{2}\delta g_{\mu\nu}\dot{x}^\mu\dot{x}^\nu + g_{\mu\nu}\dot{x}^\mu \frac{d}{ds}(\delta x^\nu)\right]ds \quad (13)$$

Now, the action for mass m at position x can be simply written as,

$$I = -\int m\, ds$$

$$\delta I = -\int [\delta(m)ds + m\delta(ds)]$$

$$= -\int \left[\frac{\partial m}{\partial x^\lambda}\delta x^\lambda + \frac{m}{2}\delta g_{\mu\nu}\dot{x}^\mu\dot{x}^\nu + m g_{\mu\nu}\dot{x}^\mu \frac{d}{ds}(\delta x^\nu)\right]ds \quad (14)$$

Integrate the last term by parts and switch dummy variable $\nu \to \lambda$ we get,

$$\delta I = -\int \left[\frac{\partial m}{\partial x^\lambda}\delta x^\lambda + \frac{m}{2}\frac{\partial g_{\mu\nu}}{\partial x^\lambda}\dot{x}^\mu\dot{x}^\nu \delta x^\lambda - \frac{d}{ds}(m g_{\mu\lambda}\dot{x}^\mu)\delta x^\lambda\right]ds$$

$$= -\int \left[\frac{\partial m}{\partial x^\lambda} + \frac{m}{2}\frac{\partial g_{\mu\nu}}{\partial x^\lambda}\dot{x}^\mu\dot{x}^\nu - \frac{d}{ds}(m g_{\mu\lambda}\dot{x}^\mu)\right]\delta x^\lambda ds = 0 \quad (15)$$

For this integral to be zero for any arbitrary δx^λ then the term in the square brackets must be zero, hence

$$\frac{d}{ds}(m g_{\mu\lambda}\dot{x}^\mu) = \frac{m}{2}\frac{\partial g_{\mu\nu}}{\partial x^\lambda}\dot{x}^\mu\dot{x}^\nu + \frac{\partial m}{\partial x^\lambda}$$

$$\frac{dm}{ds}g_{\mu\lambda}\dot{x}^\mu + m\left(g_{\mu\lambda}\frac{d\dot{x}^\mu}{ds} + g_{\mu\lambda,\nu}\dot{x}^\mu\dot{x}^\nu\right) = \frac{m}{2}g_{\mu\nu,\lambda}\dot{x}^\mu\dot{x}^\nu + \frac{\partial m}{\partial x^\lambda} \quad (16)$$

where we may make the $g_{\mu\lambda,\nu}$ term symmetric in μ,ν as follows,

$$g_{\mu\lambda}\frac{d}{ds}(m\dot{x}^\mu) = \frac{m}{2}(g_{\mu\nu,\lambda} - g_{\mu\lambda,\nu} - g_{\nu\lambda,\mu})\dot{x}^\mu\dot{x}^\nu + \frac{\partial m}{\partial x^\lambda} \qquad (17)$$

Then using the definition for the Christoffel symbol $\Gamma_{\lambda\mu,\nu}$ and multiplying throughout by $g^{\sigma\lambda}$ we get,

$$\frac{d}{ds}(m\dot{x}^\sigma) + m(g^{\sigma\lambda}\Gamma_{\lambda\mu,\nu})\dot{x}^\mu\dot{x}^\nu - g^{\sigma\lambda}\frac{\partial m}{\partial x^\lambda} = 0$$

$$\frac{d}{ds}(m\dot{x}^\sigma) + m\Gamma^\sigma_{\mu\nu}\dot{x}^\mu\dot{x}^\nu - g^{\sigma\lambda}\frac{\partial m}{\partial x^\lambda} = 0 \qquad (18)$$

Written for mass m_a at position x_a the equation of motion becomes,

$$\frac{d}{d\tau}\left(m_a\frac{dx_a^\mu}{d\tau}\right) + m_a\Gamma^\mu_{\nu\lambda}\frac{dx_a^\nu}{d\tau}\frac{dx_a^\lambda}{d\tau} - g^{\mu\nu}\frac{\partial m_a}{\partial x_a^\nu} = e_a\sum_{b\neq a}F^{(b)\mu}_\nu\frac{dx_a^\nu}{d\tau} \qquad (19)$$

where the Lorentz force has been included on the right for completeness. The world-lines of particles are not in general geodesics in the new theory. This equation agrees with the HN result in their book [12] p125 Eq.(138). In the HN book this equation of motion was derived directly from the gravitational field equation.

3. Hoyle-Narlikar Theory Development

There is some motivation for looking into the HN-theory in detail. We begin from the first of the Hoyle-Narlikar papers, through to the writing of their book. The notation they use is very unfortunate and difficult to read. There are too many similar letters being used for different parameters. Here we attempt to rewrite the theory in a more familiar notation, using Greek letters for 0,1,2,3 and roman letters only to distinguish particle "a" from particle "b". We do not use their European style of 4-vector 1,2,3,4. Rather we use the 0,1,2,3 numbering which has become fairly standard throughout the world. The flat metric will be taken as $\eta_{\mu\nu} = \text{diag}(1,-1,-1,-1)$. Where ever possible we leave c not equal to unity which helps with dimensional analysis. We tackle the papers in order starting with the first published.

Paper 1: The first paper in the sequence, in 1962, [13] was entitled "Mach's Principle and the creation of matter". The main point of the paper was to argue that although Einstein was very much influenced by Mach's ideas, he did not quite manage to get the full spirit of Mach's main idea embedded into the field equations... *mass depends on interaction with the rest of the mass-energy in the universe.*

We would argue that Mach's principle has several definitions and several of those are in fact already contained in Einstein's general relativity theory.

According to HN, they take the Einstein field equations, written as,

$$R^{\mu\nu} - \frac{1}{2}g^{\mu\nu}R + \lambda g^{\mu\nu} = -\kappa T^{\mu\nu} \tag{20}$$

and plug in the well known Robertson Walker line element,

$$ds^2 = c^2 dt^2 - S^2(t)\left[\frac{dr^2}{1-kr^2} + r^2(d\theta^2 + \sin^2\theta d\phi^2)\right] \tag{21}$$

where $k = 0, \pm 1$ and where r can be chosen for an observer attached to any particular particle. For the stress-energy tensor, they assume

$$T^{\mu\nu} = \left(\rho + P + \frac{4}{3}u\right)\frac{dx^\mu}{ds}\frac{dx^\nu}{ds} - \left(P + \frac{u}{3}\right)g^{\mu\nu} \tag{22}$$

where ρ is the matter density, P is the gas pressure and u is the radiation density. Hoyle and Narlikar set out, in a series of papers, to formulate a gravitational theory (which encompasses Einstein's equations) which included Mach's principle from the start. This theory would have both retarded and advanced waves. Essentially this would be the gravitational equivalent of Wheeler-Feynman absorber theory for electrodynamics.

Paper 2: In 1963 [14], we see HN play around with the Einstein action and add in their C-field, to add matter to the universe in an attempt to preserve the density as the universe expands. This is not really of interest for our work. Sciama [15] publishes work in the same journal on Wheeler-Feynman absorber theory and mentions Hogarth's work [2].

Paper 3: In January 1964 we see the first attempt at something new. The paper is entitled, "Time symmetric electrodynamics and the arrow of time in cosmology", [16]. Here we see the first introduction of the Fokker-Schwarzschild-Tetrode action (FST action) [17,18,7], a discussion of the Wheeler–Feynman absorber theory [1], and reworking time-symmetric electrodynamics in a flat and Riemannian space-time. Here we rewrite these familiar equations for convenience since it will set up the new notation for their later work.

We start with a summary of the first few HN equations which we then "translate" into better notation below. We quote directly from the paper [16]:

.. we consider space-time to be given by the co-ordinates x^i and by the line-element

$$ds^2 = \eta_{ik}dx^i dx^k \tag{23}$$

where η_{ik} = diag$(-1,-1,-1,+1)$. The charges are labelled by letters a, b, c.... The ath particle has co-ordinates a^i, mass m_a, charge e_a and proper time a given by

$$da^2 = \eta_{ik}da^i da^k \tag{24}$$

We have chosen the velocity of light to be unity so that the time units are the same as the space units. The Schwarzschild-Tetrode-Fokker action function is then defined by

$$J = -\sum_a m_a \int da - \sum_a \sum_{b \neq a} \frac{1}{2} e_a e_b \int \int \delta(ab_i ab^i) \eta_{lm} da^l da^m \qquad (25)$$

where

$$\delta(ab_i ab^i) = \eta_{ik}(a^i - b^i)(a^k - b^k) \qquad (26)$$

The equations of motion can be obtained from (25) by requiring that J be stationary with respect to variations of the world lines of particles. If we define the 4-potential of b at a point x by the function

$$A_m^{(b)}(x) = \int e_b \delta(xb_i xb^i) \eta_{mk} db^k \qquad (27)$$

the equations of motion take the form

$$m_a \frac{d^2 a^k}{da^2} = e_a \sum_{b \neq a} F_l^{k(b)} \frac{da^l}{da} \qquad (28)$$

where

$$F_{kl}^{(b)} = \frac{\partial A_l^{(b)}(x)}{\partial x^k} - \frac{\partial A_k^{(b)}(x)}{\partial x^l} \qquad (29)$$

represents the 'field' of charge b at point x.

Note that a better notation of the FST action and derivations for the potential and Maxwell's equations, can be found at the very end of the book by Barut on electrodynamics [19]. Our notation is similar to Barut's only we use x instead of z. Also we use a and b to distinguish particles rather than α and β since these could easily be mistaken for summation variables. We start by rewriting the above notation as follows: **For flat space-time.**

Define the metric as $\eta_{\mu\nu} = \text{diag}(+1, -1, -1, -1)$. The charges are labelled by a, b, c as before. The ath particle has coordinates x_a^μ, mass m_a, charge e_a, and proper time τ given by the line element as

$$ds^2 = c^2 d\tau^2 = dx_a^\mu dx_{a\mu} , \qquad (30)$$

with $c = 1$ we get,

$$\begin{aligned} d\tau^2 &= \eta_{\mu\nu} dx_a^\mu dx_a^\nu \\ d\tau &\to \eta_{\mu\nu} \dot{x}_a^\mu \dot{x}_a^\nu d\tau . \end{aligned} \qquad (31)$$

where differentiation w.r.t τ is represented by the dot above the symbol. The action can be written as,

$$I = -\sum_a \int \frac{1}{2} m_a (\dot{x}_a^\nu)^2 d\tau - \sum_a \sum_{b \neq a} e_a \int A_\mu^{(b)}(x_a^\nu) \dot{x}_a^\mu d\tau \tag{32}$$

$$A_\mu^{(b)}(x_a^\nu) = \int e_b D(x_a - x_b) \eta_{\mu\nu} dx_b^\nu \equiv \int e_b D(x_a - x_b) \eta_{\mu\nu} \dot{x}_b^\nu d\tau' \tag{33}$$

$$D(x_a - x_b) = \left[\eta_{\alpha\beta}(x_a^\alpha - x_b^\alpha)(x_a^\beta - x_b^\beta) \right]$$

Note that the 4-potential of particle b ($A_\mu^{(b)}$) is evaluated at the location of particle a. The proper time for particle b is given by τ' and $\dot{x}_b = dx_b/d\tau'$. The Lagrangian can be written as

$$I = \sum_a \int L(x_a^\nu, \dot{x}_a^\nu) d\tau \tag{34}$$

$$L(x_a^\nu, \dot{x}_a^\nu) = -\frac{1}{2} m_a (\dot{x}_a^\nu)^2 - e_a \sum_{b \neq a} A_\mu^{(b)}(x_a^\nu) \dot{x}_a^\mu \tag{35}$$

with equation of motion given by

$$\frac{\partial L}{\partial x_a^\nu} - \frac{d}{d\tau}\left(\frac{\partial L}{\partial \dot{x}_a^\nu}\right) = 0 \ . \tag{36}$$

Differentiating the Lagrangian we get

$$\frac{\partial L}{\partial x_a^\nu} = -\sum_{b \neq a} e_a \frac{\partial A_\mu^{(b)}}{\partial x_a^\nu} \dot{x}_a^\mu$$

$$\frac{\partial L}{\partial \dot{x}_a^\nu} = -m_a \dot{x}_a^\nu - \sum_{b \neq a} e_a A_\nu^{(b)}$$

$$\frac{d}{d\tau}\left(\frac{\partial L}{\partial \dot{x}_a^\nu}\right) = -\frac{d}{d\tau}(m_a \dot{x}_a^\nu) - e_a \frac{\partial A_\nu^{(b)}}{\partial x_a^\mu} \frac{dx_a^\mu}{d\tau}$$

$$m_a \ddot{x}_a^\nu = e_a \dot{x}_a^\mu \sum_{b \neq a} F_{\nu\mu}^{(b)}(x_a^\nu) \tag{37}$$

where the last equation is the equation of motion of particle a. The mass m_a is taken to be constant. Finally we define the electromagnetic field tensor as

$$F_{\nu\mu}^{(b)}(x_a^\nu) = \left(\frac{\partial A_\mu^{(b)}}{\partial x_a^\nu} - \frac{\partial A_\nu^{(b)}}{\partial x_a^\mu}\right)$$

$$F_{\nu\mu}^{(b)} = \frac{1}{2}\left(F_{\nu\mu}^{(b)\text{ret}} + F_{\nu\mu}^{(b)\text{adv}}\right) \ . \tag{38}$$

Now we follow the paper but write only in the new notation. Using Dirac's identity,

$$\eta^{\mu\nu}\frac{\partial^2}{\partial x^\mu \partial x^\nu}D(x-x_b) = -4\pi\delta(x^0-x_b^0)(x^1-x_b^1)(x^2-x_b^2)(x^3-x_b^3)$$
$$= -4\pi\delta^4(x-x_b) \ . \tag{39}$$

The 4-potential satisfies [19],

$$\Box^2 A^{(b)\mu}(x) = \sum_{b\neq a} e_b \int \dot{x}_b^\mu(\tau') \Box^2 D(x-x_b) d\tau'$$
$$= -4\pi \sum_{b\neq a} e_b \int \dot{x}_b^\mu \delta^4(x-x_b) d\tau' \tag{40}$$

which is the same as writing,

$$\Box^2 A^{(b)\mu}(x) = \eta^{\mu\nu}\frac{\partial^2}{\partial x^\mu \partial x^\nu} A_\sigma^{(b)}(x) = -4\pi j_\sigma^{(b)}(x) \tag{41}$$

where the current density $j_\sigma^{(b)}(x)$ is given by

$$j_\sigma^{(b)}(x) = e_b \int_{-\infty}^{\infty} \eta_{\sigma\lambda}\dot{x}_b^\lambda \delta^4(x-x_b) d\tau' \tag{42}$$

It can be shown that [19], the 4-potential satisfies the gauge condition

$$\frac{\partial A^{(b)\mu}}{\partial x^\mu} = \sum_{b\neq a} e_b \int \dot{x}_b^\mu \frac{\partial}{\partial x^\mu} D(x-x_b) d\tau'$$
$$= -D(x-x_b) \mid_{\tau'=-\infty}^{+\infty} = 0 \ . \tag{43}$$

We may derive the following:

$$\frac{\partial F_\nu^{(b)\mu}}{\partial x^\mu} = -4\pi j_\nu^{(b)}(x) \ , \tag{44}$$

which are Maxwell's inhomogeneous equations.

Formally, all Maxwell's equations and the Lorentz force equation are derivable from the action principle, except radiation reaction terms (self force terms). The radiation reaction becomes a force due to advanced waves coming from the absorbing universe mass-energy. The time symmetry is emphasized by rewriting the 4-potential as a sum of retarded and advanced parts.

$$A_\mu^{(b)} = \frac{1}{2}\left(A_\mu^{(b)\text{Ret}} + A_\mu^{(b)\text{Adv}}\right)$$
$$A_\mu^{(b)\text{Ret}} = e_b \frac{\eta_{\mu\nu}\dot{x}_b^\nu}{\eta_{\alpha\beta}(x^\alpha-x_b^\alpha)\dot{x}^\beta} \tag{45}$$

An alternative approach to the first term in the Lagrangian is to is to vary it directly and derive the equations of motion from scratch rather than using the Euler-Lagrange formula. The notation has now been introduced so we will not continue with the Riemannian Space-time summary.

Paper 4 & 5: These two papers [20,21] are referring entirely to the C-field, which was an addition of matter in order that the mass-density of the universe ρ remain constant as the universe expands. We are not interested in the C-field, since we do not require a static universe, and wish to treat the universe as not only expanding but accelerating in that expansion. We skip these two papers.

Paper 6, 7, & 8: Now we jump ahead to the full HN-theory and the fully Machian action, or in their words, the *full action*. The first of these [222] is a short paper including the C-field. This is not of so much interest. The next two papers [3,4] are the main papers with the theory we wish to use. These two papers should be read together. The HN-theory is given in *A new theory of gravitation* 1964 [3] with extra details in the 1966 paper [4] entitled *A conformal theory of gravitation*.

A summary of the new theory [3] follows with reference also to the extra detail in [4]. Particle interactions are propagated along null geodesics (at no distance in a 4 dimensional or light-like sense). According to HN the action developed thus far is of the form

$$I = \frac{1}{16\pi G} \int R\sqrt{(-g)}d^4x - \sum_a m_a \int d\tau - \sum_a \sum_{b\neq a} 4\pi e_a e_b \int\int G_{\alpha\beta} dx_a^\alpha dx_b^\beta \tag{46}$$

the first two terms looking very different that the direct particle interaction representing the electromagnetic last term. The term in m_a is derived from Galileo's concept of inertia and has been present since before Newton. Einstein retained this traditional term. Neither of the first two terms is correct, the first being a field or energy density the second being attributed to matter only. Only terms of the form using a double integral should be present. The first two terms have been artificially separated by traditional thinking. In what follows we construct a purely gravitational theory with the first and second terms combined into a single mass-energy term. It may also be possible to combine the electromagnetic term into the same term but we leave that for a later discussion. In order to convert the line integral $\int m_a d\tau$ into a sum of double line integrals we make the following assumptions:

(1) The mass $m_a = m(x_a)$ (mass at position x_a) must become a direct particle field, it must arise from all the other mass in the universe.

(2) Since mass is scalar we expect it to arise through a scalar Greens function.

(3) The action must be symmetric between any pairs of particles, [3].

Let each particle b give rise to a mass-field (*spherical monopole type g-waves*). Denote this field at a general point x by $m^{(b)}(x)$. At any point x_a on the path of particle a, we have $m^{(b)}(x_a)$ as the contribution of particle b to the mass of particle a at the position x_a. Summing for all b particles,

$$m(x_a) = m_a = \sum_b m^{(b)}(x_a) \qquad (47)$$

this gives the mass at point x_a due to all particles including those at position x_a. For electromagnetism we avoided positions where $x_a = x_b$ but for gravity we need not do this, [12] p109 Eq(46). The non-electromagnetic part of the action I_{mat} for many particles a,b... can be written in the form,

$$I_{\text{mat}} = -\frac{1}{2}\sum_a \int m(x_a)d\tau = -\sum_a \sum_b \int m^{(b)}(x_a)d\tau \qquad (48)$$

In order that (48) be symmetric for any particle pair a,b we must have $m^{(b)}(x_a)$ in the form

$$m^{(b)}(x_a) = \int G(x_a, x_b)d\tau' \qquad (49)$$

so that

$$\int m^{(b)}(x_a)d\tau = \int\int G(x_a, x_b)d\tau d\tau' \qquad (50)$$

where $G(x_a, x_b) = G(x_b, x_a)$ is a scalar Greens function. The mass function at a point x due to the world-line of particle a, at position x_a, is defined by

$$m^{(a)}(x) = \int G(x, x_a)d\tau \ . \qquad (51)$$

The mass function varies from point to point. *Before we plunge into the depths of HN-theory, let us first have a brief aside on the development of the field equations for the Einstein action, which is considerably easier!*

4. Comparison of the Actions and Field equations.

4.1 The Einstein Action

For comparison we write down the basic Einstein Action, without the electromagnetic field,

$$I_{\text{Einstein}} = \frac{1}{16\pi G}\int R[-g]^{1/2}d^4x - \sum_a m_a \int d\tau \ . \qquad (52)$$

The field equations are derived by varying the action and setting the variation equal to zero, [12] p112. The metric tensor will be varied according to $g_{\mu\nu} \to g_{\mu\nu} + \delta g_{\mu\nu}$ in a volume V with $\delta g_{\mu\nu} = 0$ at the boundaries. Varying the above action yields,

$$\begin{aligned}\delta I_{\text{Einstein}} &= \frac{1}{16\pi G}\int \delta(R[-g]^{1/2})d^4x - \sum_a m_a \int \delta(d\tau) \\ &= \frac{1}{16\pi G}\int \delta(g^{\mu\nu}R_{\mu\nu}[-g]^{1/2})d^4x - \sum_a m_a \int \delta(d\tau) \quad .(53)\end{aligned}$$

Using

$$\begin{aligned}-\sum_a m_a \int \delta(d\tau) &= \frac{1}{2}\int T_{\mu\nu}\delta g^{\mu\nu}[-g]^{1/2}d^4x \\ \delta\left([-g]^{1/2}\right) &= -\frac{1}{2}g_{\mu\nu}\delta g^{\mu\nu}[-g]^{1/2} \quad (54)\end{aligned}$$

Next we expand out the first term with the Ricci tensor,

$$\begin{aligned}\delta I_{\text{Einstein}} &= \frac{1}{16\pi G}\int \delta(g^{\mu\nu}R_{\mu\nu})[-g]^{1/2}d^4x \\ &\quad - \frac{1}{16\pi G}\int \frac{1}{2}Rg_{\mu\nu}\delta g^{\mu\nu}[-g]^{1/2}d^4x \\ &\quad + \frac{1}{2}\int T_{\mu\nu}\delta g^{\mu\nu}[-g]^{1/2}d^4x \\ &= \frac{1}{16\pi G}\int \left[R_{\mu\nu} - \frac{1}{2}Rg_{\mu\nu} + 8\pi G T_{\mu\nu}\right]\delta g^{\mu\nu}[-g]^{1/2}d^4x \\ &\quad + \frac{1}{16\pi G}\int g^{\mu\nu}\delta R_{\mu\nu}[-g]^{1/2}d^4x \quad (55)\end{aligned}$$

the last term is zero since the variation vanishes on the boundary. Hence by setting $\delta I_{\text{Einstein}} = 0$ for any arbitrary variation $\delta g_{\mu\nu}$ we obtain the Einstein's field equations,

$$R_{\mu\nu} - \frac{1}{2}Rg_{\mu\nu} = -8\pi G T_{\mu\nu} \quad (56)$$

The familiar energy momentum tensor is easily derived, see Hoyle and Narlikar's book [12] p112. This follows from using $d\tau^2 = g_{\mu\nu}dx^\mu dx^\nu$ and thus $\delta(d\tau) = \delta g_{\mu\nu}\dot{x}^\mu \dot{x}^\nu d\tau$ which leads to

$$\begin{aligned}-\sum_a \int m(x_a)\delta(d\tau) &= -\sum_a \int m(x)\delta g_{\mu\nu}\dot{x}^\mu \dot{x}^\nu \delta^4(x-x_a)d\tau \\ &= -\int_V T^{\mu\nu}\delta g_{\mu\nu}[-g]^{1/2}d^4x \\ &= +\int_V T_{\mu\nu}\delta g^{\mu\nu}[-g]^{1/2}d^4x \\ \text{where } T^{\mu\nu} &= \sum_a \delta^4(x-x_a)[-g]^{-1/2}m(x)\dot{x}^\mu \dot{x}^\nu d\tau \quad (57)\end{aligned}$$

The energy-stress tensor $T^{\mu\nu}$ is a sum over all the mass-energy in the universe, excluding the electromagnetic field which is treated separately. This is exactly the same calculation that will appear in the HN-theory later.

4.2 Quick note on scalar densities

Using J as the Jacobian (see Dirac's book on gravitation [11] p37),

$$\begin{aligned} dx^{\mu'} &= dx^{\mu} J \quad \text{or} \quad d^4x' = J d^4x \\ J &= \frac{\partial x^{\mu'}}{\partial x^{\alpha}} \\ g_{\alpha\beta} &= \frac{\partial x^{\mu'}}{\partial x^{\alpha}} g_{\mu'\nu'} \frac{\partial x^{\nu'}}{\partial x^{\beta}} \end{aligned} \quad (58)$$

The determinants satisfy,

$$\begin{aligned} g &= J g' J \\ g &= J^2 g' \\ \Rightarrow \sqrt{-g} &= J \sqrt{-g'} \end{aligned} \quad (59)$$

since $g = \|g_{\alpha\beta}\|$ is negative. That makes $\sqrt{-g}$ a positive quantity. Hence we may define the following invariant quantity for any scalar density, for example $H \to T_{\mu\nu} \delta g^{\mu\nu}$,

$$\int_V H \sqrt{-g} d^4x = \int_V H \sqrt{-g'} J d^4x = \int_V H' \sqrt{-g'} d^4x'$$

hence $\int_V T_{\mu\nu} \delta g^{\mu\nu} \sqrt{-g} d^4x =$ invariant . $\quad (60)$

4.3 The HN-Theory Action

Omitting the electromagnetic field for now, using the definitions (47) and (50), the action can be written, following Hoyle-Narlikar "A New Theory of Gravitation", [3], as:

$$I = -\sum_a \frac{1}{2} \int m(x_a) d\tau = -\sum_a \sum_b \int \int G(x_a, x_b) d\tau d\tau' \quad (61)$$

There is just one term, a sum over all the masses in the universe. The energy is not separated out, because of mass-energy equivalence. This requires that a "universe" consist of at least two particles for them to interact and create a space-time between them. The factor $1/2$ comes in because each $G(x_a, x_b)$ is shared by two particles a and b. This makes no difference to the equations of motion. The paper has no factor of $1/2$

in front of the double sum, whereas the HN book does have the factor of 1/2. The most general wave equation is

$$g^{\mu\nu}G(x,x_a)_{;\mu\nu} + \mu R G(x,x_a) = -[-g]^{-1/2}\delta^4(x-x_a) \quad (62)$$

in which R is the Ricci scalar and μ is a constant taken later to be 1/6 since the wave equation is then conformally invariant [23]. The next step is to vary the geometry in a finite volume V, $g_{\mu\nu} \to g_{\mu\nu} + \delta g_{\mu\nu}$ with $\delta g_{\mu\nu} = 0$ on the boundary. It will be shown that

$$\delta I = \int P_{\mu\nu}\delta g^{\mu\nu}[-g]^{1/2}d^4x \quad (63)$$

where $P_{\mu\nu}$ is a symmetric tensor. The formalism becomes a theory when we assert that $\delta I = 0$ which requires

$$P_{\mu\nu} = 0 \quad (64)$$

which are the field equations of the new theory.

4.4 Field equation for HN-theory

Now for the field equations, [3]. Consider the change in $G(x_a, x_b)$ due to an infinitesimal change $\delta g_{\mu\nu}$ in $g_{\mu\nu}$ over a finite volume V, with $\delta g_{\mu\nu} = 0$ on the boundary of V. By dividing throughout by $[-g]^{-1/2}$, the equation for the Greens function $G(x, x_a)$ can be written as,

$$\frac{\partial}{\partial x^\mu}\left[[-g]^{1/2}g^{\mu\nu}\frac{\partial G(x,x_a)}{\partial x^\nu}\right] + \mu R[-g]^{1/2}G(x,x_a) = -\delta^4(x-x_a) \quad (65)$$

The variation can be made by setting $G \to G + \delta G$ and $g^{\mu\nu} \to g^{\mu\nu} + \delta g^{\mu\nu}$, and this becomes

$$\begin{aligned}&\frac{\partial}{\partial x^\mu}\left[[-g]^{1/2}g^{\mu\nu}\frac{\partial \delta G}{\partial x^\nu}\right] + \mu R[-g]^{1/2}\delta G \\ &= -\frac{\partial}{\partial x^\mu}\left[\delta([-g]^{1/2}g^{\mu\nu})\frac{\partial G}{\partial x^\nu}\right] - \mu\delta(R[-g]^{1/2})G \end{aligned} \quad (66)$$

This agrees with Eq (71) in the HN book, [12] p113-114. It appears that δG satisfies the same differential operator as $G(x, x_a)$ itself, but with a distributed source term, not a δ-function at point x_a. The solution for δG can be written down as follows, (see [26] for first use of this solution on the scalar Greens function p186)

$$\delta G(x_a, x_b) = \int_V \frac{\partial}{\partial x^\mu} \left[\delta([-g]^{1/2} g^{\mu\nu}) \frac{\partial G(x_a, x)}{\partial x^\nu} \right] G(x_b, x) d^4x$$

$$+ \mu \int_V \delta(R[-g]^{1/2}) G(x_a, x) G(x_b, x) d^4x$$

$$= - \int_V \delta([-g]^{1/2} g^{\mu\nu}) \frac{\partial G(x_a, x)}{\partial x^\nu} \frac{\partial G(x_b, x)}{\partial x^\mu} d^4x$$

$$+ \mu \int_V \delta(R[-g]^{1/2}) G(x_a, x) G(x_b, x) d^4x \quad (67)$$

where we have integrated the first term by parts and set $\delta g^{\mu\nu} = 0$ at the boundary of the volume. This agrees with Eq (12) in [3] (and Eq (72) in the HN book). The variation of the action then becomes

$$\delta I = -\frac{1}{2} \sum_a \int m(x_a) \delta(d\tau) - \frac{1}{2} \sum_a \int \delta m(x_a) d\tau$$

$$= -\frac{1}{2} \sum_a \int m(x_a) \delta(d\tau) - \sum_a \sum_b \int \int \delta G(x_a, x_b) d\tau d\tau'$$

$$= -\frac{1}{2} \sum_a \int m(x_a) \delta(d\tau)$$

$$- \sum_a \sum_b \int_V \int \int \delta([-g]^{1/2} g^{\mu\nu}) \frac{\partial G(x_a, x)}{\partial x^\nu} \frac{\partial G(x_b, x)}{\partial x^\mu} d^4x d\tau d\tau'$$

$$+ \mu \sum_a \sum_b \int_V \int \int \delta(R[-g]^{1/2}) G(x_a, x) G(x_b, x) d^4x d\tau d\tau' \quad (68)$$

Using the earlier definitions of mass at point x due to the world-lines of particle a and particle b, Eq.(236),

$$m^{(a)}(x) = \int G(x_a, x) d\tau$$

$$m^{(b)}(x) = \int G(x_b, x) d\tau'$$

we arrive at

$$\delta I = -\frac{1}{2} \sum_a \int m(x_a) \delta(d\tau)$$

$$- \sum_a \sum_b \int_V \delta([-g]^{1/2} g^{\mu\nu}) \frac{\partial m^{(a)}(x)}{\partial x^\nu} \frac{\partial m^{(b)}(x)}{\partial x^\mu} d^4x$$

$$+ \mu \sum_a \sum_b \int_V \delta(R[-g]^{1/2}) m^{(a)}(x) m^{(b)}(x) d^4x \quad (69)$$

This agrees with Eq (13) in [3]. There are typos in the papers making these look like covariant derivatives when they are only partial derivatives.

The first term in the variation of the action, Eq. (69) is the familiar energy momentum tensor for mass-energy. This follows from using $d\tau^2 = g_{\mu\nu}dx^\mu dx^\nu$:

$$
\begin{aligned}
-\sum_a \int m(x_a)\delta(d\tau) &= -\sum_a \int m(x)\delta g_{\mu\nu}\dot{x}^\mu \dot{x}^\nu \delta^4(x-x_a)d\tau \\
&= -\int_V T^{\mu\nu}\delta g_{\mu\nu}[-g]^{1/2}d^4x \\
&= \int_V T_{\mu\nu}\delta g^{\mu\nu}[-g]^{1/2}d^4x \\
\text{where } T^{\mu\nu} &= \sum_a \delta^4(x-x_a)[-g]^{-1/2}m(x)\dot{x}^\mu \dot{x}^\nu d\tau \quad (70)
\end{aligned}
$$

This is exactly the same as for the Einstein action treated earlier. This does not include the electromagnetic fields which are treated separately.

At this point rather than follow the paper [3], it appeared quicker to follow the book [12]. We take up the derivation there. In order to compare the older paper [3] with the more recent text book [12] we return to the variation of the Greens function Eq (66). The book uses $\mu = 1/6$ and has a factor of $1/2$ in front of the double sum, so the following terms will have a multiplicative factor $1/2$ throughout. We may split the Green function into advanced and retarded parts, [12] p114 Eq (73),

$$G(x,x_b) = \frac{1}{2}\left[G^{\text{ret}}(x,x_b) + G^{\text{adv}}(x,x_b)\right] . \quad (71)$$

The retarded part gives the following contribution to $\delta G(x_a, x_b)$, see earlier Eq (67),

$$
\begin{aligned}
\delta G^{\text{ret}}(x_a, x_b) = &-\frac{1}{2}\int_V G^{\text{ret}}(x_a, x)\frac{\partial}{\partial x^\mu}\left[\delta([-g]^{1/2}g^{\mu\nu})\frac{\partial G^{\text{ret}}(x,x_b)}{\partial x^\nu}\right]d^4x \\
&-\frac{1}{12}\int_V \delta(R[-g]^{1/2})G^{\text{ret}}(x_a,x)G^{\text{ret}}(x,x_b)d^4x \quad (72)
\end{aligned}
$$

where $G^{\text{ret}}(x_a, x) = G^{\text{adv}}(x, x_a)$.

The equation for δG^{ret} above, can be written more symmetrically by integrating the first term by parts,

$$
\begin{aligned}
\delta G^{\text{ret}} = &\frac{1}{2}\int_V \delta([-g]^{1/2}g^{\mu\nu})\frac{\partial G^{\text{adv}}(x,x_a)}{\partial x^\mu}\frac{\partial G^{\text{ret}}(x,x_b)}{\partial x^\nu}d^4x \\
&-\frac{1}{12}\int_V \delta(R[-g]^{1/2})G^{\text{adv}}(x,x_a)G^{\text{ret}}(x,x_b)d^4x
\end{aligned}
$$
(73)

This agrees with the book [12] p115, Eq. (77). The advanced part of δG is similar with the advanced and retarded G's switched

$$\begin{aligned}\delta G^{\text{adv}} &= \frac{1}{2}\int_V \delta([-g]^{1/2}g^{\mu\nu})\frac{\partial G^{\text{ret}}(x,x_a)}{\partial x^\mu}\frac{\partial G^{\text{adv}}(x,x_b)}{\partial x^\nu}d^4x \\ &\quad -\frac{1}{12}\int_V \delta(R[-g]^{1/2})G^{\text{ret}}(x,x_a)G^{\text{adv}}(x,x_b)d^4x\end{aligned}$$
(74)

The full expression for $\delta G(x_a, x_b)$ is the sum of the advanced and retarded parts. The next step is to find the variation of the action,

$$\sum_a \sum_b \int\int \delta G(x_a, x_b)d\tau d\tau' .$$
(75)

Here we introduce the mass field from p115 [112],

$$\begin{aligned} m(x) &= \frac{1}{2}[m^{(\text{ret})}(x) + m^{(\text{adv})}(x)] \\ m^{(\text{ret})}(x) &= \sum_a \int G^{(\text{ret})}(x,x_a)d\tau \\ m^{(\text{adv})}(x) &= \sum_a \int G^{(\text{adv})}(x,x_a)d\tau \end{aligned}$$
(76)

The variation of G becomes,

$$\begin{aligned}&\sum_a\sum_b \int\int \delta G(x_a,x_b)d\tau d\tau' \\ =\;& \frac{1}{2}\sum_a\sum_b\int_V\int\int \delta([-g]^{1/2}g^{\mu\nu})\left[\frac{\partial G^{\text{adv}}(x,x_a)}{\partial x^\mu}\frac{\partial G^{\text{ret}}(x,x_b)}{\partial x^\nu}\right.\\ +\;& \left.\frac{\partial G^{\text{ret}}(x,x_a)}{\partial x^\mu}\frac{\partial G^{\text{adv}}(x,x_b)}{\partial x^\nu}\right]d^4x d\tau d\tau' \\ -\;& \frac{1}{12}\sum_a\sum_b\int_V\int\int \delta(R[-g]^{1/2})\left[G^{\text{adv}}(x,x_a)G^{\text{ret}}(x,x_b)\right.\\ +\;& \left.G^{\text{ret}}(x,x_a)G^{\text{adv}}(x,x_b)\right]d^4x d\tau d\tau'\end{aligned}$$
(77)

Note that each term has one sum over a and one over b, so when we substitute in the mass fields all the summations are used.

$$\sum_a \sum_b \int \int \delta G(x_a, x_b) d\tau d\tau'$$
$$= \frac{1}{2} \int_V \delta([-g]^{1/2} g^{\mu\nu}) \left[\frac{\partial m^{\text{adv}}}{\partial x^\mu} \frac{\partial m^{\text{ret}}}{\partial x^\nu} + \frac{\partial m^{\text{ret}}}{\partial x^\mu} \frac{\partial m^{\text{adv}}}{\partial x^\nu} \right] d^4x$$
$$- \frac{1}{12} \int_V \delta(R[-g]^{1/2}) \left[m^{\text{adv}} m^{\text{ret}} + m^{\text{ret}} m^{\text{adv}} \right] d^4x \tag{78}$$

We may therefore simplify the δI to remove the summations. Here is a full summary so far:

$$\begin{aligned}
\delta I &= -\frac{1}{2} \delta \left[\sum_a \int m(x_a) d\tau \right] \\
&= -\frac{1}{2} \sum_a m(x_a) \delta(d\tau) - \frac{1}{2} \sum_a \int \delta m(x_a) d\tau \\
&= -\frac{1}{2} \int_V T^{\mu\nu} \delta g_{\mu\nu} [-g]^{1/2} d^4x - \frac{1}{2} \sum_a \sum_b \int \int \delta G(x_a, x_b) d\tau d\tau' \\
&= +\frac{1}{2} \int_V T_{\mu\nu} \delta g^{\mu\nu} [-g]^{1/2} d^4x \\
&\quad - \frac{1}{2} \int_V \delta([-g]^{1/2} g^{\mu\nu}) \left[\frac{\partial m^{\text{adv}}}{\partial x^\mu} \frac{\partial m^{\text{ret}}}{\partial x^\nu} \right] d^4x \\
&\quad + \frac{1}{12} \int_V \delta(R[-g]^{1/2}) \left[m^{\text{adv}} m^{\text{ret}} \right] d^4x
\end{aligned} \tag{79}$$

where we have replaced the first term with the familiar energy-stress tensor expression and flipped from contravariant to covariant notation with a minus sign change. The second term in the above equation can be expanded to give;

$$-\frac{1}{2} \delta([-g]^{1/2} g^{\mu\nu}) \frac{\partial m^{\text{adv}}}{\partial x^\mu} \frac{\partial m^{\text{ret}}}{\partial x^\nu} =$$
$$-\frac{1}{2} \left[\delta[-g]^{1/2} g^{\alpha\beta} \frac{\partial m^{\text{adv}}}{\partial x^\alpha} \frac{\partial m^{\text{ret}}}{\partial x^\beta} + [-g]^{1/2} \delta g^{\mu\nu} \frac{\partial m^{\text{adv}}}{\partial x^\mu} \frac{\partial m^{\text{ret}}}{\partial x^\nu} \right]$$
$$= \frac{1}{2} \left[\left(\frac{1}{2} g_{\mu\nu} \delta g^{\mu\nu} [-g]^{1/2} \right) g^{\alpha\beta} \frac{\partial m^{\text{adv}}}{\partial x^\alpha} \frac{\partial m^{\text{ret}}}{\partial x^\beta} \right]$$
$$- [-g]^{1/2} \delta g^{\mu\nu} \frac{1}{2} \left[\frac{\partial m^{\text{adv}}}{\partial x^\mu} \frac{\partial m^{\text{ret}}}{\partial x^\nu} + \frac{\partial m^{\text{adv}}}{\partial x^\nu} \frac{\partial m^{\text{ret}}}{\partial x^\mu} \right] \tag{80}$$

where we have used the following useful identity [12] p 113, in the last step

$$\delta[-g]^{1/2} = -\frac{1}{2} g_{\mu\nu} \delta g^{\mu\nu} [-g]^{1/2} \tag{81}$$

The $\delta(R[-g]^{1/2})$ term in δI can be expanded also as follows;

$$\frac{1}{6}\delta(R[-g]^{1/2}) = \frac{1}{6}\delta(R_{\mu\nu}g^{\mu\nu}[-g]^{1/2})$$
$$= \frac{1}{6}\left[\delta(g^{\mu\nu}R_{\mu\nu})[-g]^{1/2} + R\delta([-g]^{1/2})\right] \quad (82)$$

using the Eq (81) again for the last term we find,

$$\frac{1}{6}\delta(R[-g]^{1/2})$$
$$= \frac{1}{6}\left[R_{\mu\nu}[-g]^{1/2}\delta g^{\mu\nu} + g^{\mu\nu}\delta R_{\mu\nu}[-g]^{1/2} - \frac{1}{2}Rg_{\mu\nu}[-g]^{1/2}\delta g^{\mu\nu}\right]$$
$$= \frac{1}{6}\left[\left(R_{\mu\nu} - \frac{1}{2}Rg_{\mu\nu}\right)\delta g^{\mu\nu}[-g]^{1/2} + g^{\mu\nu}\delta R_{\mu\nu}[-g]^{1/2}\right] \quad (83)$$

hence the contribution to δI becomes

$$\frac{1}{6}\int_V \delta(R[-g]^{1/2})\left[m^{\mathrm{adv}}m^{\mathrm{ret}}\right]d^4x$$
$$= \frac{1}{6}\int_V \left(R_{\mu\nu} - \frac{1}{2}Rg_{\mu\nu}\right)\delta g^{\mu\nu}[-g]^{1/2}\left[m^{\mathrm{adv}}m^{\mathrm{ret}}\right]$$
$$+ \frac{1}{6}\int_V g^{\mu\nu}\delta R_{\mu\nu}[-g]^{1/2}\left[m^{\mathrm{adv}}m^{\mathrm{ret}}\right]d^4x \quad (84)$$

We will treat the $\delta R_{\mu\nu}$ term separately, it does not go to zero as in the Einstein case *unfortunately!*

$$\frac{1}{12}\int_V g^{\mu\nu}\delta R_{\mu\nu}[-g]^{1/2}\left[m^{\mathrm{adv}}m^{\mathrm{ret}}\right]d^4x = \frac{1}{2}\int \theta_{\mu\nu}\delta g^{\mu\nu}[-g]^{1/2}d^4x$$
$$\theta_{\mu\nu} = \frac{1}{6}\frac{g^{\mu\nu}}{\delta g^{\mu\nu}}\delta R_{\mu\nu}m^{\mathrm{adv}}m^{\mathrm{ret}} \quad (85)$$

The shorthand for δI then becomes,

$$\delta I = +\frac{1}{2}\int_V T_{\mu\nu}\delta g^{\mu\nu}[-g]^{1/2}d^4x$$
$$+\frac{1}{2}\left[\left(\frac{1}{2}g_{\mu\nu}\delta g^{\mu\nu}[-g]^{1/2}\right)g^{\alpha\beta}\frac{\partial m^{\mathrm{adv}}}{\partial x^\alpha}\frac{\partial m^{\mathrm{ret}}}{\partial x^\beta}\right]$$
$$-[-g]^{1/2}\delta g^{\mu\nu}\frac{1}{2}\left[\frac{\partial m^{\mathrm{adv}}}{\partial x^\mu}\frac{\partial m^{\mathrm{ret}}}{\partial x^\nu} + \frac{\partial m^{\mathrm{adv}}}{\partial x^\nu}\frac{\partial m^{\mathrm{ret}}}{\partial x^\mu}\right]$$
$$+\frac{1}{6}\int_V \left(R_{\mu\nu} - \frac{1}{2}Rg_{\mu\nu}\right)\delta g^{\mu\nu}[-g]^{1/2}\left[m^{\mathrm{adv}}m^{\mathrm{ret}}\right]$$
$$+\frac{1}{2}\int \theta_{\mu\nu}\delta g^{\mu\nu}[-g]^{1/2}d^4x = 0 . \quad (86)$$

The field equations are then seen to be,

$$T_{\mu\nu} + \theta_{\mu\nu} + \frac{1}{6}(R_{\mu\nu} - \frac{1}{2}g_{\mu\nu}R)m^{\text{adv}}m^{\text{ret}}$$
$$- \frac{1}{2}\left[\frac{\partial m^{\text{adv}}}{\partial x^\mu}\frac{\partial m^{\text{ret}}}{\partial x^\nu} + \frac{\partial m^{\text{adv}}}{\partial x^\nu}\frac{\partial m^{\text{ret}}}{\partial x^\mu} - g_{\mu\nu}g^{\alpha\beta}\frac{\partial m^{\text{adv}}}{\partial x^\alpha}\frac{\partial m^{\text{ret}}}{\partial x^\beta}\right] = 0 \tag{87}$$

What remains is to expand $\theta_{\mu\nu}$ in its full glory. See the Addendum for details.

After some trivial algebra, which is obvious to the most casual observer, and only takes a couple of pages of calculation we get...

$$\theta_{\mu\nu} = -\frac{1}{6}\left[g_{\mu\nu}\Box^2(m^{\text{adv}}m^{\text{ret}}) - (m^{\text{adv}}m^{\text{ret}})_{,\mu\nu}\right] \tag{88}$$

where $\Box^2$ is the wave equation $\partial_\mu \partial^\mu$.

5. Derivation of Woodward's Mass change formula

5.1 Woodward's Power Equation → mass change formula

From Woodward's book [24], page 73 Eq(3.5), we find;

$$\delta m = \frac{1}{4\pi G}\left[\frac{1}{\rho_0 c^2}\frac{\partial P}{\partial t} - \left(\frac{1}{\rho_0 c^2}\right)^2 \frac{P^2}{V}\right]$$

$$\delta m = \frac{1}{4\pi G}\left[\frac{V}{m_0 c^2}\frac{\partial^2 \varepsilon}{\partial t^2} - \left(\frac{V}{m_0 c^2}\right)^2 \frac{1}{V}\left(\frac{\partial \varepsilon}{\partial t}\right)^2\right]$$

$$= \frac{1}{4\pi G}\left[\frac{V}{m_0}\frac{\partial^2 m}{\partial t^2} - V\left(\frac{1}{m_0}\right)^2\left(\frac{\partial m}{\partial t}\right)^2\right]$$

$$\frac{\delta m}{V} = \frac{1}{4\pi G}\left[\frac{1}{m_0}\frac{\partial^2 m}{\partial t^2} - \left(\frac{1}{m_0}\right)^2\left(\frac{\partial m}{\partial t}\right)^2\right] \tag{89}$$

where V is volume over the device, P is power to the device, and $P = d\varepsilon/dt$. Energy is $\varepsilon = mc^2$ and mass density $\rho_0 = m_0/V$. This agrees with the dimensions of $[G] = [FL^2/M^2]$.

5.2 HN-theory field equation → mass change formula

Let's define the HN-field equation (in a smooth fluid) as follows (which agrees with Eq.(16) in reference [4]) by grouping terms together;

$$R_{\alpha\beta} - \frac{1}{2}g_{\alpha\beta}R = -8\pi G(T_{\alpha\beta} + \delta T_{\alpha\beta}) \quad \text{where}$$

$$-(8\pi G)\delta T_{\alpha\beta} = \frac{2}{m}(g_{\alpha\beta}g^{\mu\nu}m_{;\mu\nu} - m_{;\alpha\beta})$$

$$+ \frac{4}{m^2}(m_{;\alpha}m_{;\beta} - \frac{1}{4}m_{;\gamma}m^{;\gamma}g_{\alpha\beta})$$

(90)

Now we expand the terms out. Let us put back in c and not set it equal to one, which can be confusing. The terms in μ, ν mix the time and spatial derivatives in an unexpected way.

Consider first the T_{00} and T_{jj} terms separately, using flat metric (+1,-1,-1,-1),

$$\begin{aligned}
-\frac{8\pi G}{c^4}\delta T_{00} &= \frac{2}{m}\left[g_{00}\left(\frac{g^{00}}{c^2}\frac{\partial^2 m}{\partial t^2} + g^{jj}\frac{\partial^2 m}{\partial x_j^2}\right) - \frac{1}{c^2}\frac{\partial^2 m}{\partial t^2}\right] \\
&\quad + \frac{4}{m^2}\left[\frac{1}{c^2}\left(\frac{\partial m}{\partial t}\right)^2 - \frac{g_{00}}{4}\left(\frac{1}{c^2}\left(\frac{\partial m}{\partial t}\right)^2 - \left(\frac{\partial m}{\partial x_j}\right)^2\right)\right] \\
&= \frac{2}{m}\left[\frac{1}{c^2}\frac{\partial^2 m}{\partial t^2} - \frac{\partial^2 m}{\partial x_j^2} - \frac{1}{c^2}\frac{\partial^2 m}{\partial t^2}\right] \\
&\quad + \frac{1}{m^2}\left[\frac{4}{c^2}\left(\frac{\partial m}{\partial t}\right)^2 - \frac{1}{c^2}\left(\frac{\partial m}{\partial t}\right)^2 + \left(\frac{\partial m}{\partial x_j}\right)^2\right] \\
&= -\frac{2}{m}\frac{\partial^2 m}{\partial x_j^2} + \frac{1}{m^2}\left(\frac{\partial m}{\partial x_j}\right)^2 + \frac{3}{m^2 c^2}\left(\frac{\partial m}{\partial t}\right)^2
\end{aligned}$$

(91)

where we treat the derivatives with respect to $\partial/\partial x_j$ as a 3 component gradient-like term.

$$\begin{aligned}
-\frac{8\pi G}{c^4}\delta T_{jj} &= \frac{2}{m}\left[g_{jj}\left(\frac{g^{00}}{c^2}\frac{\partial^2 m}{\partial t^2}+g^{jj}\frac{\partial^2 m}{\partial x_j^2}\right)-\frac{\partial^2 m}{\partial x_j^2}\right]\\
&\quad +\frac{4}{m^2}\left[\left(\frac{\partial m}{\partial x_j}\right)^2-\frac{g_{jj}}{4}\left(\frac{1}{c^2}\left(\frac{\partial m}{\partial t}\right)^2-\left(\frac{\partial m}{\partial x_j}\right)^2\right)\right]\\
&= \frac{2}{m}\left[-\frac{1}{c^2}\frac{\partial^2 m}{\partial t^2}+\frac{\partial^2 \cancel{m}}{\cancel{\partial x_j^2}}-\frac{\partial^2 \cancel{m}}{\cancel{\partial x_j^2}}\right]\\
&\quad +\frac{1}{m^2}\left[4\left(\frac{\partial m}{\partial x_j}\right)^2+\frac{1}{c^2}\left(\frac{\partial m}{\partial t}\right)^2-\left(\frac{\partial m}{\partial x_j}\right)^2\right]\\
&= -\frac{2}{mc^2}\frac{\partial^2 m}{\partial t^2}+\frac{3}{m^2}\left(\frac{\partial m}{\partial x_j}\right)^2+\frac{1}{m^2c^2}\left(\frac{\partial m}{\partial t}\right)^2
\end{aligned}$$
(92)

where $j=1,2$ or 3. Now take the trace of $T_{\alpha\alpha}$ where $\alpha=0,1,2,3$ by adding the last two equations.

$$\begin{aligned}
-\frac{8\pi G}{c^4}\text{Tr}(\delta T_{\alpha\alpha}) &= -\frac{2}{m}\frac{\partial^2 m}{\partial x_j^2}+\frac{4}{m^2}\left(\frac{\partial m}{\partial x_j}\right)^2+\frac{4}{m^2 c^2}\left(\frac{\partial m}{\partial t}\right)^2\\
&\quad -\frac{2}{mc}\frac{\partial^2 m}{\partial t^2}\\
\frac{1}{c^2}\text{Tr}(\delta T_{\alpha\alpha}) &= \frac{1}{4\pi G}\left[\left\{\frac{1}{m}\frac{\partial^2 m}{\partial t^2}-\frac{2}{m^2}\left(\frac{\partial m}{\partial t}\right)^2\right\}\right.\\
&\quad \left.+\left\{\frac{c^2}{m}\frac{\partial^2 m}{\partial x_j^2}-\frac{2c^2}{m^2}\left(\frac{\partial m}{\partial x_j}\right)^2\right\}\right]
\end{aligned}$$
(93)

where we assume we are summing over α and j. This last expression should be compared with the previous result Eq. (89) above. Note that there are also spatial terms here, which in previous papers I incorporated into the time derivatives [28]. I now think that was a mistake and have written them out explicitly here. This is the main result of the paper. Quoting from a paper [29] by R. Medina:

> "Unlike the inertia of energy, which is well known, many physicists are not aware of the inertia of pressure (stress). In many cases such an effect is negligible, but for the case of the stress produced by electrostatic interactions, it is comparable to the inertial effects of the electromagnetic fields. If the inertia of stress is neglected the calculations are inconsistent."

The spatial and temporal terms may be related, in the sense that in The Mach effect drive (or MEGA drive), PZT (lead zirconate titanate) expands and contracts. In a different device, that may not be the case.

6. Conclusions

The main result we wish to emphasize is the mass fluctuation, Eq. (93). Compare this with Woodward's result Eq. (89) from his book [24] p73, Eq. (3.5). The consequences of this mass fluctuation are *astounding* as related to the Woodward effect and propellant-less propulsion methods. A following paper in this chapter, by Rodal, will describe how to calculate a force using the mass fluctuation calculated here. The calculated force and resonant frequency predictions will be compared to experimental data.

Hoyle-Narlikar gravitation, or gravitational absorber theory (GAT), is a valid theory that is fully consistent with Einstein's GR. It is a fully Machian theory of gravitation, which means that the mass of a body depends on its gravitational interaction with all the other masses in the universe. Text books on Einstein's GR rarely if ever mention advanced waves, yet the are necessary if interactions with distant matter are to be thought of as instantaneous.

Around 1965 Hawking voiced an objection to HN-theory [27], but that objection is no longer valid due to the accelerating expansion of the universe [28]. Hoyle-Narlikar theory is to gravitation what Wheeler-Feynman absorber theory is to electromagnetism. Einstein's General Relativity (GR) remains valid and all the tests of Einstein's GR also remain valid and carry over to the HN theory presented here. The real difference is in the highly symmetric and simplified Lagrangian, which treats a mass as being influenced by all the other masses in the universe, and that is all. A real universe must therefore be made up of at least two masses.

Acknowledgements

HF thanks Lance Williams for helpful suggestions and for clarification of the introduction section. HF also thanks José Rodal for the suggestion of using the name *gravitational absorber theory* (GAT) as opposed to HN theory, to try and distance the basic theory from the ideas of static universe cosmology. Rodal also points out that GAT could stand for *gravitationally assisted trajectory* and I noticed it also could stand for *glorious awe-inspiring theory*.

Addendum

For those of you who just can't get enough algebra, here is the rest of the glorious details for the derivation of $\theta_{\mu\nu}$.

We need to expand out Eq. (85) and find the equation for $\theta_{\mu\nu}$;

$$\frac{1}{2}\int_V \theta_{\mu\nu}\delta g^{\mu\nu}[-g]^{1/2}d^4x = \frac{1}{12}\int_V (\delta R_{\mu\nu}g^{\mu\nu})m^{\text{adv}}m^{\text{ret}}[-g]^{1/2}d^4x$$

The term in the round bracket on the RHS of the equation can be written as, [12], p118 Eqs (98,99).

$$(g^{\mu\nu}\delta R_{\mu\nu}) = \frac{1}{[-g]^{1/2}}\frac{\partial}{\partial x^\lambda}\left[[-g]^{1/2}w^\lambda\right]$$
$$w^\lambda = (g^{\mu\nu}\delta\Gamma^\lambda_{\mu\nu} - g^{\mu\nu}\delta\Gamma^\nu_{\mu\nu}) \tag{94}$$

Hence we may write,

$$\frac{1}{12}\int_V (\delta R_{\mu\nu}g^{\mu\nu})m^{\text{adv}}m^{\text{ret}}[-g]^{1/2}d^4x$$
$$= \frac{1}{12}\int_V \frac{\partial}{\partial x^\lambda}\left[[-g]^{1/2}w^\lambda\right](m^{\text{adv}}m^{\text{ret}})d^4x$$
$$= \frac{1}{12}\int_V w^\lambda \frac{\partial}{\partial x^\lambda}(m^{\text{adv}}m^{\text{ret}})[-g]^{1/2}d^4x \tag{95}$$

where we have integrated by parts. This agrees with Eq (100) in Hoyle and Narlikar's book [12]. Using the following identities, from their book p118, w^λ can be expanded.

$$\Gamma^\nu_{\mu\nu} = \frac{1}{[-g]^{1/2}}\frac{\partial}{\partial x^\mu}\left([-g]^{1/2}\right)$$
$$g^{\mu\nu}\Gamma^\lambda_{\mu\nu} = -\frac{1}{[-g]^{1/2}}\frac{\partial}{\partial x^\nu}\left([-g]^{1/2}g^{\lambda\nu}\right)$$
$$\delta([-g]^{-1/2}) = +\frac{1}{2}g_{\mu\nu}\delta g^{\mu\nu}[-g]^{-1/2}$$
$$\frac{\partial}{\partial x^\alpha}[-g]^{1/2} = [-g]^{1/2}\Gamma^\beta_{\alpha\beta} \tag{96}$$

we will also be reusing the identity in Eq (81), which is also in reference [11] p50, Eq. (26.10).

Consider the following,

$$\delta(g^{\mu\nu}\Gamma^\lambda_{\mu\nu}) = \delta g^{\mu\nu}\Gamma^\lambda_{\mu\nu} + g^{\mu\nu}\delta\Gamma^\lambda_{\mu\nu}$$
$$\Rightarrow w^\lambda = (g^{\mu\nu}\delta\Gamma^\lambda_{\mu\nu} - g^{\mu\nu}\delta\Gamma^\nu_{\mu\nu})$$
$$= \delta(g^{\mu\nu}\Gamma^\lambda_{\mu\nu}) - \delta g^{\mu\nu}\Gamma^\lambda_{\mu\nu} - g^{\mu\nu}\delta\Gamma^\nu_{\mu\nu} \tag{97}$$

Now we need to consider the separate parts of the equation for w^λ and rewrite it;

$$\delta\Gamma^\nu_{\mu\nu} = \delta([-g]^{-1/2})\frac{\partial}{\partial x^\mu}([-g]^{1/2}) + ([-g]^{-1/2})\frac{\partial}{\partial x^\mu}(\delta[-g]^{1/2})$$

$$\delta(g^{\mu\nu}\Gamma^\lambda_{\mu\nu}) = -\delta([-g]^{-1/2})\frac{\partial}{\partial x^\nu}([-g]^{1/2}g^{\lambda\nu})$$

$$- [-g]^{-1/2}\frac{\partial}{\partial x^\nu}[\delta([-g]^{1/2}g^{\lambda\nu})] \quad (98)$$

where we have differentiated the identities (96) above. Now we substitute these expressions into the equation for w^λ to obtain,

$$w^\lambda = -\delta g^{\mu\nu}\Gamma^\lambda_{\mu\nu} - \frac{1}{[-g]^{1/2}}\frac{\partial}{\partial x^\nu}[\delta([-g]^{1/2}g^{\lambda\nu})]$$

$$- \frac{1}{[-g]^{1/2}}g^{\mu\lambda}\frac{\partial}{\partial x^\mu}[\delta[-g]^{1/2}]$$

$$- g^{\alpha\lambda}\delta([-g]^{-1/2})\left[\frac{\partial}{\partial x^\alpha}[-g]^{1/2}\right]$$

$$+ \delta([-g]^{-1/2})[-g]^{1/2}g^{\alpha\beta}\Gamma^\lambda_{\alpha\beta} \quad (99)$$

Now we only need to substitute the following identities,

$$\delta([-g]^{1/2}) = -\frac{1}{2}g_{\mu\nu}\delta g^{\mu\nu}[-g]^{1/2}$$

$$\delta([-g]^{-1/2}) = +\frac{1}{2}g_{\mu\nu}\delta g^{\mu\nu}[-g]^{-1/2}$$

$$\frac{\partial}{\partial x^\alpha}[-g]^{1/2} = [-g]^{1/2}\Gamma^\beta_{\alpha\beta} \quad (100)$$

to obtain the needed result for w^λ,

$$w^\lambda = -\delta g^{\mu\nu}\Gamma^\lambda_{\mu\nu} - \frac{1}{[-g]^{1/2}}\frac{\partial}{\partial x^\nu}\left[\delta([-g]^{1/2}g^{\lambda\nu})\right]$$

$$- \frac{1}{[-g]^{1/2}}g^{\mu\lambda}\frac{\partial}{\partial x^\mu}\left[\delta[-g]^{1/2}\right]$$

$$- \frac{1}{2}g^{\alpha\lambda}\Gamma^\beta_{\alpha\beta}g_{\mu\nu}\delta g^{\mu\nu} + \frac{1}{2}\delta g^{\mu\nu}g_{\mu\nu}g^{\alpha\beta}\Gamma^\lambda_{\alpha\beta} \quad (101)$$

which agrees with Eq (102) p118 [12]. Now at this point, this wonderful expression for w^λ must be placed back inside the integral (276), because we need to find the result for $\theta_{\mu\nu}$. The three terms involving Christoffel symbols cancel out. You can integrate by parts and use the divergence theorem. The only remaining terms involve differentiations on the mass functions only.

$$\frac{1}{2}\int \theta_{\mu\nu}\delta g^{\mu\nu}[-g]^{1/2}d^4x = -\frac{1}{6}\int \left(-\frac{1}{[-g]^{1/2}}\frac{\partial}{\partial x^\alpha}\left[\delta([-g]^{1/2}g^{\lambda\alpha})\right] - \right.$$
$$\left.\frac{1}{[-g]^{1/2}}g^{\alpha\lambda}\frac{\partial}{\partial x^\alpha}\left[\delta[-g]^{1/2}\right]\right)$$
$$\times \frac{\partial}{\partial x^\lambda}(m^{\text{adv}}m^{\text{ret}})[-g]^{1/2}d^4x \quad (102)$$

Expand out the first term and substitute for the
$\delta[-g]^{1/2} = -\frac{1}{2}g_{\mu\nu}\delta g^{\mu\nu}[-g]^{1/2}$, then the $[-g]^{1/2}$ terms cancel out:

$$\Rightarrow \theta_{\mu\nu}\delta g^{\mu\nu} = \left(-\frac{1}{12}g^{\lambda\alpha}g_{\mu\nu}\delta g^{\mu\nu}\frac{\partial}{\partial x^\alpha} - \frac{1}{12}g^{\alpha\lambda}g_{\mu\nu}\delta g^{\mu\nu}\frac{\partial}{\partial x^\alpha}\right)$$
$$\frac{\partial}{\partial x^\lambda}(m^{\text{adv}}m^{\text{ret}})$$
$$+\frac{1}{6}\delta g^{\lambda\alpha}\frac{\partial}{\partial x^\alpha}\frac{\partial}{\partial x^\lambda}(m^{\text{adv}}m^{\text{ret}}) \quad (103)$$

Now we must substitute $\delta g^{\lambda\alpha} \to \delta g^{\mu\nu}$ to match the LHS of the equation in the last term. Performing contractions over α on the first two terms, leads to,

$$\theta_{\mu\nu} = -\frac{1}{6}\left(g_{\mu\nu}\frac{\partial^2}{\partial x^\lambda \partial x_\lambda}(m^{\text{adv}}m^{\text{ret}}) - \frac{\partial^2}{\partial x^\mu \partial x^\nu}(m^{\text{adv}}m^{\text{ret}})\right)$$
$$\text{thus } \theta_{\mu\nu} = -\frac{1}{6}\left(g_{\mu\nu}\Box^2(m^{\text{adv}}m^{\text{ret}}) - (m^{\text{adv}}m^{\text{ret}})_{,\mu\nu}\right).$$
$$(104)$$

References

[1] J. A. Wheeler and R. P., Feynman, "Interaction with the absorber as a mechanism of radiation", Rev. Mod. Phys. **17** 157 (1945).

[2] J. E. Hogarth, " Considerations of the Absorber Theory of Radiation", Proc. Roy. Soc. **A267**, pp365-383 (1962).

[3] F. Hoyle and J. V. Narlikar, "A new theory of gravitation", Proc. Roy. Soc. Lon. **A282**, 191 (1964).

[4] F. Hoyle and J. V. Narlikar, "A Conformal theory of gravitation", Proc. Roy. Soc. Lon. **A294**, 138 (1966).

[5] P. A. M. Dirac, "Classical theory of radiating electrons", Proc. Roy. Soc. Lon. **A167**, 148 (1938).

[6] J. D. Jackson, "*Classical Electrodynamics*", 3rd edition, John Wiley and Sons, Inc. p771, (1999).

[7] H. Tetrode, Zeits. fur Physik, **10**, 317 (1922).

[8] G. N. Lewis, Nat. Acad. Sci. Proc. **12**, 22 (1926).

[9] Cramer, J. G. "The transactional Interpretation of quantum mechanics." Rev. Mod. Phys. **58**, 647 (1986)

[10] Cramer, J. G. *The Quantum Handshake, Entanglement, Nonlocality and Transactions.* Springer, Berlin (2015)

[11] P. A. M. Dirac, *"General Theory of Relativity"*, Princeton Landmarks in Physics, (Princeton Univ. Press 1996.) p 16 for geodesic derivation.

[12] F. Hoyle and J. V. Narlikar, *"Action at a distance in Physics and Cosmology"* W. H. Freeman and Company, San Francisco, 1974.

[13] F. Hoyle and J. V. Narlikar, "Mach's Principle and the Creation of Matter", Proc. Roy. Soc. Lon. **A270**, pp334-341 (1962).

[14] F. Hoyle and J. V. Narlikar, "Mach's principle and the creation of matter", Proc. Roy. Soc. Lon. **A273**, pp1-11 (1963).

[15] Sciama, D. W. , "Retarded potentials and the expansion of the universe", Proc. Roy. Soc. Lond. **A273**, pp484-495 (1963).

[16] F. Hoyle and J. V. Narlikar, "Time Symmetric Electrodynamics and the Arrow of Time in Cosmology", Proc. Roy. Soc. Lon. **A277**, pp 1-23 (1964).

[17] Fokker, A. D. Zeit. fur Physik, **58**, 386 (1929).

[18] Schwarzschild K., Nachr. Ges. Wiss. Gottingen **128**, 132 (1903).

[19] A.O. Barut, " *Electrodynamics and classical ttheory of fields and particles"*, pp213-220 (Dover Publications Inc. New York 1964).

[20] F. Hoyle and J. V. Narlikar, "On the Avoidance of Singularities in C-field Cosmology", Proc. Roy. Socv. Lon. **A278**, pp465-478 (1964).

[21] F. Hoyle and J. V. Narlikar, "The C-field as a Direct Particle Field", Proc. Roy. Soc. Lon. **A282**, pp 178-183 (1964).

[22] F. Hoyle and J. V. Narlikar, "On the gravitational influence of direct particle fields", Proc. Roy. Soc. Lon. **A282**, 184 (1964). See also "A conformal theory of gravitation", Proc. Roy. Soc. Lon. **A294**, 138 (1966).

[23] B. S. deWitt and R. W. Brehme, Annals of Physis **9**, 220 (1960).

[24] J. F. Woodward, *Making Starships and Stargates*, Springer Press, December 2012.

[25] H. Fearn et al. 'Theory of a Mach Effect Thruster II", Journal of Modern Physics, **6**, pp1868–1880 (2015).

[26] Rodrigo Medina, "The inertia of Stress", Am. J. Phys. **74** 1031 (2006).
https://arxiv.org/pdf/physics/0609144.pdf

[27] S. W. Hawking, "On the Hoyl-Narlikar Theory of Gravitation", Proc. Roy. Soc. Lon. **A 286**, 313 (1965).

[28] H. Fearn, "Mach's Principle, Action at a Distance and Cosmology", Journal of Modern Physics, **6**, pp260–272 (2015).

Verification of the Thrust Signature of a Mach Effect Device

Nembo Buldrini
FOTEC Forschungs und Technologietransfer GmbH
2700 Wiener Neustadt, Austria

A Mach Effect Thruster is an apparatus based on piezoelectric material, which is supposed to produce thrust via an interaction of its components with, chiefly, the distant mass of the universe. A device of this sort, built and tested by Woodward, has been tested on a thrust balance in high vacuum at FOTEC (Austria). The results confirm qualitatively the presence of the same effect observed by Woodward.

1. Introduction

Being able to reach another planetary system in a reasonable amount time would only be possible if some kind of propellantless propulsion were developed. Propellantless systems like laser sails are an option, but they come with the drawback of relying on an external and distant source of power. The alternative of a self-contained propellantless system, capable of producing movement with seemingly no interaction with its surroundings, is considered to be the ultimate space propulsion scheme.

Since 1990, James F. Woodward and collaborators have shown that it should be possible, via Mach's Principle, to achieve such a scheme [1,2]. If the theory developed by Woodward were verified and validated experimentally, it would not only have practical consequences for space flight, but would also tell us more about the structure of our universe, as it would shed light on the origin of a fundamental property of mass, inertia. The last experimental embodiment of this theory is a device named MET, Mach Effect Thruster. The same device is also known as MEGA, Mach Effect Gravity Assist, an acronym which better describes the underlying working principle.

Figure 1: The MET tested at FOTEC (lid of the Faraday cage removed)

A device of this sort has been sent to the author by Woodward in 2014. The device (Figure 1) is constituted by a stack of piezoelectric discs (material: lead zirconium titanate) clamped between an aluminium cap (left in the figure) and a brass reaction mass, and mounted inside an aluminium Faraday cage lined with mu-metal foil.

Applying a sinusoidal voltage to the piezoelectric stack causes it to change in size and shape, according to its piezoelectric and electrostrictive properties. The combined effect of these deformations is thought to produce thrust by an exchange of momentum via gravitational interactions with the distant cosmic masses [1,2].

What follows is a description of the tests this device underwent at FOTEC during the spring of 2014.

2. Experimental Setup

The device has been installed on a thrust balance which has been developed to measure the thrust produced by liquid metal ion thrusters, usually ranging from some μN to more than 1 mN. The balance is of the torsion type, with vertical rotation axis, and the pivot consisting of two flexural bearings. The deflection of the balance is detected by a fiber optic displacement sensor. More details on the balance construction and verification can be found in [3].

The electrical connections to the device are implemented via a stack of liquid metal (Galinstan) contacts placed at the pivot: this method assures virtually no friction and no spurious forces produced when power is fed through the contacts (at reasonably low current/voltage values).

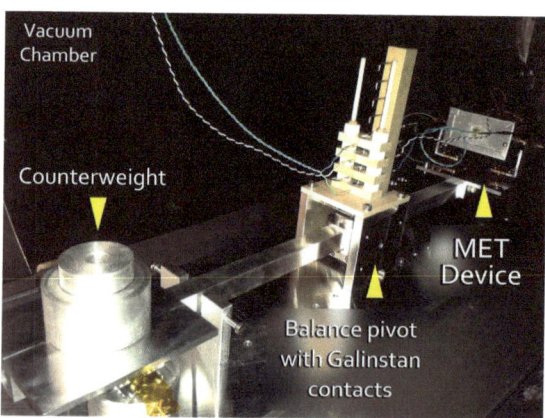

Figure 2: The MET device mounted on the thrust balance inside the vacuum chamber

The device, mounted on the thrust balance, has been tested inside a vacuum chamber (Figure 2). The vacuum chamber is a cylindrical steel chamber of about 800 mm in diameter and 1750 mm in length; it is equipped with a roughing pump and a turbo pump, and pressures as low as $10^{-6} - 10^{-7}$ mbar ($10^{-4} - 10^{-5}$ Pa) can be customarily achieved.

Preliminary testing has been performed with only two electric lines to the device: the power line, which provided power to the stack of piezoelectric disks, and the temperature measurement line, for the measurement of the temperature on the aluminium cap end. Between the two available temperature measurement sensors, located respectively at the aluminium cap end and at brass reaction mass end, the first one has been chosen because of the faster response due to the lower thermal mass.

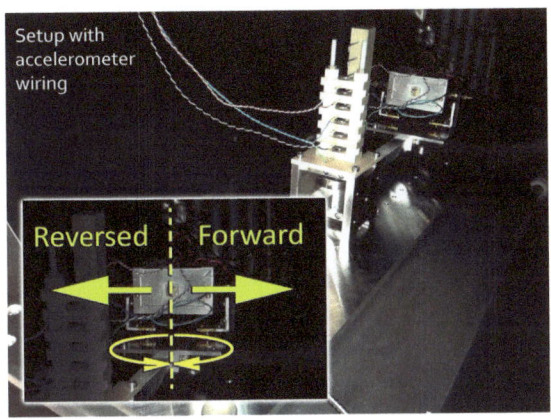

Figure 3: Setup with added accelerometer wiring and explanation of the thrust vector direction. When going from forward to reversed thrust, the device is rotated around the axis represented by the dashed yellow line.

In subsequent testing, a third line has been added for measuring the voltage across a thin piezoelectric disk, part of the stack, which is passively used as an accelerometer (Figure 3). This enabled better monitoring of the operating conditions of the device. If fact, in order for Mach effects to be produced, both piezoelectric and electrostrictive force components must be present, and this can be confirmed by the presence of second harmonic content in the accelerometer signal (Figure 4).

The power to the device has been supplied by a Carvin DCM 2500 amplifier, through a step-up transformer, capable of increasing the voltage of the amplifier to levels suitable for a proper operation of the piezoelectric device. Both the amplifier and the step-up transformer have the same specifications like the ones used by Woodward.

3. Test Results

All the tests have been run when the pressure in the vacuum chamber was 3×10^{-6} mbar (3×10^{-4} Pa) or lower. When a sinusoidal voltage

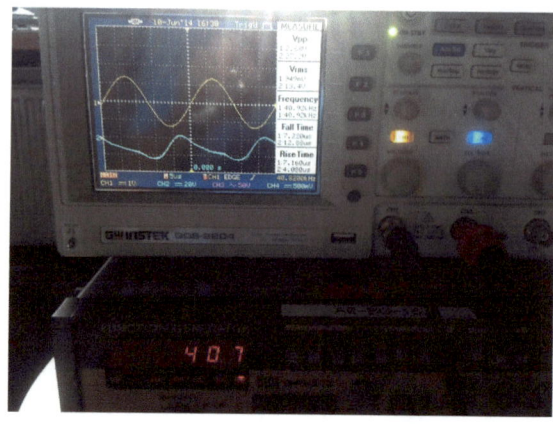

Figure 4: Setup with added accelerometer wiring and explanation of the thrust vector direction. When going from forward to reversed thrust, the device is rotated around the axis represented by the dashed yellow line.

of about 200 V peak to peak with a frequency of about 40 kHz is applied to the device, a thrust signal of about 0.15 μN is produced, with reliable repeatability, provided the temperature of the device was in the range between 30°C and 60°C. The chosen operating frequency of 40 kHz corresponds to the maximum thrust production, and it has been selected after testing the device at different frequency values.

The plots displayed in Figure 5 and Figure 8 show the direct response of the balance at the activation of the device. No further elaboration has been applied at the data. The gray bands indicate the time when power is supplied to the device.

Figure 6 compares the trace in Figure 5 with a trace obtained by Woodward when operating a twin device at approximately the same power level and frequency. Although the thrust magnitude is different, in both graphs a distinctive pattern can be identified, characterized by the presence of transitory effects occurring at the start and at the end of the operating time. Three phases can be recognized: (1) a starting transient constituted by a peak going in a direction opposite to the following steady thrust, (2) a steady thrust period of about 5 seconds, (3) at switch-off, a peak going in the same direction of the thrust. Figure 7 depicts an interpretation of the structure of the signal, where noise, drift and part of the overshooting are removed for sake of clarity.

Figure 8 shows a series of three runs in a row of the device after this has been rotated by 180° (direction reversal), as indicated in Fig. 3. Figure 8 serves also to point out the good repeatability of the thrust signal.

The fact that we have this distinctive common thrust pattern, which keeps its overall shape consistently across different devices of the same

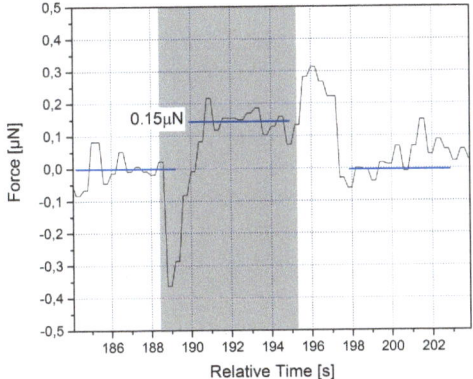

Figure 5: Typical thrust plot. The area in gray indicates the time when the device is operating

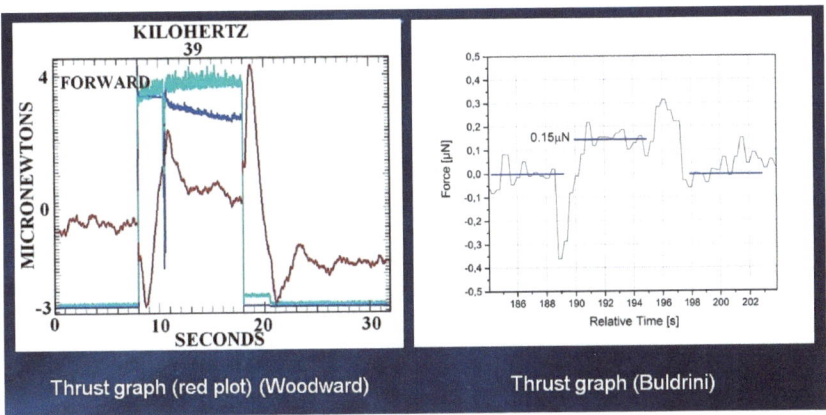

Figure 6: Comparison with a typical MET thrust signature obtained by Woodward

sort and different testing setups[1], and reverses with 180° rotation of the device, without changing in magnitude, is a strong indication that the effect is originating from the device itself and not from an interaction of the device with the close surroundings, nor from some interaction between different parts of the setup. In addition, if this signature is characteristic of the device, it may offer important clues on its actual operation, and could be compared with the results of models which try to characterize Mach effects in this kind of devices, like the one presented by J. Rodal in

---

[1] To date, two other replications (by G. Hathaway and M. Tajmar) have reported comparable results

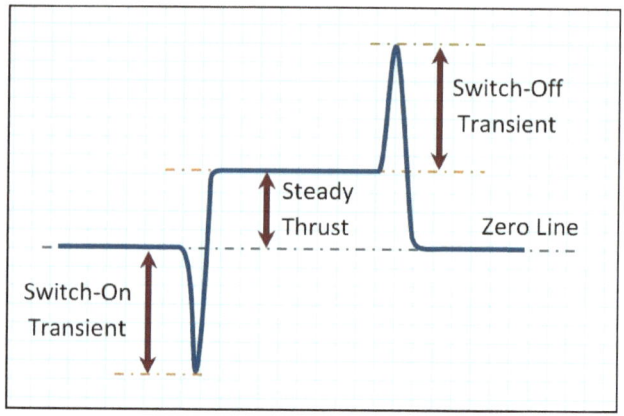

Figure 7: Structure of the detected thrust signal.

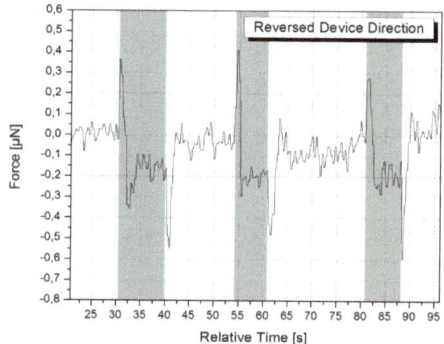

Figure 8: Thrust plots with the device positioned in reversed direction.

these workshop proceedings.

Figure 9 makes a comparison of the plot in Figure 5 with a thrust plot obtained from the same device operated by Woodward at similar power level and frequency. It is interesting to note here that, while the transients are clearly visible and larger when compared with the measurement obtained at FOTEC, the steady thrust is difficult if not impossible to discern, due to what it seems a combination of noise, drift and zero-line offset.

In general, the different magnitude of the steady thrust value across otherwise similar and similarly operated devices could be ascribed to a sum of factors, these being, for example, degradation of device components (piezo stacks are sensitive to moisture), slight constructional differences between tested devices, different effective power delivered to the

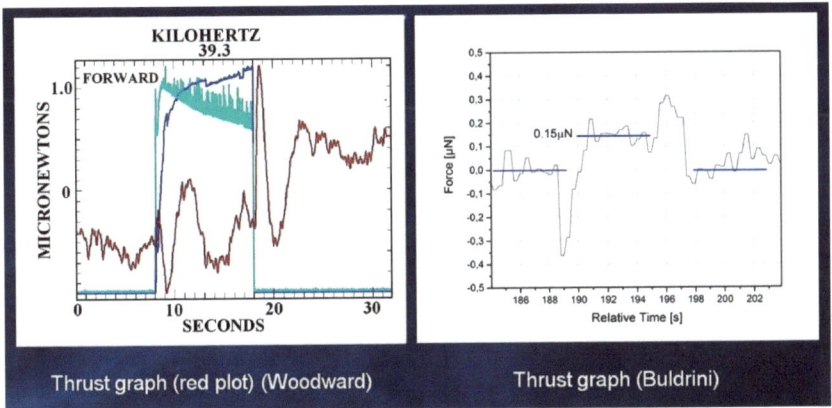

Figure 9: Comparison of the thrust signature obtained from the same device by Woodward (left) and the author (right).

device and balance calibration issues. An additional factor, which would explain the mismatch in the magnitude of the transients, can be a different moment of inertia of the balances. Lighter balance beams, in fact, would react faster, showing larger transient peaks.

4. Conclusions

While previous replication efforts by the author on a different type of Mach effects device (Mach Lorenz Thruster) have produced inconclusive results [4], the kind of signal produced by the MET device here reported corroborates the results obtained by Woodward.

The distinctive shape of the thrust signal and its reversal with the rotation of the device allow to assert with high confidence that the effect is taking place inside the device itself, and it is not due to some sort of interaction between different parts of the experimental setup. Taking into account the small magnitude of the effect recorded to date, the possibility that other not yet considered complex (yet mundane) effects may be in play originating the signal seen cannot be totally excluded. However, the results obtained until now, together with several null tests performed (for example [5]), permit to reduce a lot the number of possible false positives and allow to focus on the device itself and its operation.

Considering the implications that the reality of Mach effects would have in many theoretical fields, for example in cosmology, and the immense benefits that propellantless propulsion would bring to space flight, further and extensive testing and characterization of this sort of devices is highly recommended.

5. Acknowledgements

The author would like to thank Prof. Jim Woodward for providing the tested device and continued support. Thanks go also to friends and colleagues, in particular to Paul March, Bruce Long, Duncan Cumming, David Mathes and Heidi Fearn for the first insights on the preliminary results of this work, and to Alexander Reissner, Bernhard Seifert, Florin Plesescu and Thomas Hörbe for supporting the experimental activity at FOTEC.

References

[1] H. Fearn, N. van Rossum, K. Wanser and J. F. Woodward, "Theory of a Mach Effect Thruster II", Journal of Modern Physics, Volume 6, 2015, pp. 1868-1880

[2] H. Fearn, J. F. Woodward and N. van Rossum, "New Theoretical Results for the Mach Effect Thruster", 51st AIAA/SAE/ASEE Joint Propulsion Conference, AIAA Propulsion and Energy Forum, (AIAA 2015-4082)

[3] B. Seifert, A. Reissner, N. Buldrini, F. Plesescu and C. Scharlemann, "Development and Verification of a μN Thrust Balance for High Voltage Electric Propulsion Systems", The 33rd International Electric Propulsion Conference, The George Washington University, USA, October 6 –10, 2013

[4] N. Buldrini, M. Tajmar, K. Marhold and B. Seifert, "Experimental Results of the Woodward Effect on a μN Thrust Balance", AIAA-2006-4911

[5] H. Fearn, and J. Woodward, "Experimental Null Test of a Mach Effect Thruster", Journal of Space Exploration, Volume 2, Issue 2, 2013, pp. 98-105

A Conventional Post-Newtonian Mach Effect

L.L. Williams
Konfluence Research Institute
Manitou Springs, Colorado, USA

Realizing that Jim has derived a post-Newtonian field equation (equation 5 of Estes Park Quick Study VI), and that standard GR provides a prescription for generating post-Newtonian field equations, I wanted to do a straightforward calculation in GR to see what the equation corresponding to (VI-5) would be. This approach does not require Jim's ansatzes 1, 2, or 3. It departs merely from textbook results in GR.

So let us start with the field equations. In the linear theory of GR, the metric $g_{\mu\nu}$ is decomposed into a Minkowski piece $\eta_{\mu\nu}$ and a small perturbation: $g_{\mu\nu} = \eta_{\mu\nu} + h_{\mu\nu}$, where $h_{\mu\nu} \ll 1$.

Standard texts give the field equations for the linear theory in a convenient gauge. Weinberg's expression (10.1.10) in harmonic gauge is:

$$\frac{\partial^2 h_{\mu\nu}}{c^2 \partial t^2} - \nabla^2 h_{\mu\nu} = \frac{16\pi G}{c^4} \left(T_{\mu\nu} - \frac{1}{2} \eta_{\mu\nu} T^\alpha_\alpha \right) \qquad (1)$$

The Newtonian limit is recovered from the linear theory by considering the time-time (0-0) component, which relates h_{00} to $T_{00} \sim \rho c^2$. This would seem to be the reasonable departure point to connect to (VI-5). The difficulty is to recover the peculiar mass time derivatives in (VI-5). We see that the standard linear equation (1) has no time derivatives of mass, but instead time derivatives of the field.

The interaction between matter and fields is not solely described by the field equations, which describe how fields arise from matter. We also need to look at the equations of motion, which describe how fields influence matter. The equation of motion for GR is called the geodesic equation, and it describes how matter is influenced by gravity. It describes the precession of the perihelion of Mercury, for example. It is in terms of the 4-velocity $U^\mu \equiv dx^\mu/d\tau = p^\mu/m$:

$$U^\alpha \nabla_\alpha p^\mu = U^\alpha \left(\frac{\partial p^\mu}{\partial x^\alpha} + \Gamma^\mu_{\alpha\beta} p^\beta \right) = \frac{dp^\mu}{d\tau} + \Gamma^\mu_{\alpha\beta} U^\alpha p^\beta = 0 \qquad (2)$$

Equation (2), along with the Einstein field equations, is the whole content of GR. The non-relativistic, post-Newtonian limits of (2) are given by Weinberg in section 3.4, and by Schutz in section 7.2. In this case, the spatial components of the 4-momentum p^μ are assumed much less than the time component. Then the non-relativistic limit of the geodesic equation (2) is:

$$\frac{dp^\mu}{d\tau} + \Gamma^\mu_{00} U^0 p^0 = Order(v/c) \tag{3}$$

The linear affine connection is given by

$$\Gamma^\mu_{00} \simeq \frac{1}{2}\eta^{\mu\nu}\left(2\frac{\partial h_{0\nu}}{c\partial t} - \frac{\partial h_{00}}{\partial x^\nu}\right) \tag{4}$$

Returning to the field equations, consider a simple stress-energy tensor, for a cold fluid or dust: $T_{\mu\nu} = \rho W_\mu W_\nu$, where $\int \rho W^\alpha dV \equiv \Sigma p^\alpha$ of the particles. Then

$$T^\alpha_\alpha \simeq \rho \left(\frac{cdt}{d\tau}\right)^2 = \rho W_0^2 \tag{5}$$

where again we ignore the spatial components of the 4-momentum relative to the time component, this time now for the bulk matter.

Let us put this together, setting $h_{00} \equiv -2\phi/c^2$, to accord with the usual identification of the Newtonian potential ϕ with the time-time component of the metric perturbation. Then the linear field equation (1) can be written:

$$\nabla^2 \phi - \frac{\partial^2 \phi}{c^2 \partial t^2} = \frac{4\pi G}{c^2} \rho W_0^2 \tag{6}$$

Likewise, the linear equation of motion (3) for the time component p^0 of the 4-momentum is (cf. also Carroll 7.23):

$$\frac{U^0}{c}\frac{\partial p^0}{\partial t} = \frac{U^0}{c^3}\frac{\partial \phi}{\partial t}p^0 \tag{7}$$

Equation (7) is typically dropped in textbook treatments of linearized gravity, on the assumption that the Newtonian limit should be time-independent. Those treatments are typically interested in the spatial pieces of (3) that show trajectories under gravitational influence. Allowing this time dependence is a nice insight that Jim has brought to this discussion.

Equation (7) allows us to write the time derivative of the field in terms of particle energy:

$$\frac{\partial^2 \phi}{\partial t^2} = \frac{\partial}{\partial t}\left(\frac{c^2}{p^0}\frac{\partial p^0}{\partial t}\right) = \frac{c^2}{p^0}\frac{\partial^2 p^0}{\partial t^2} - \left(\frac{c}{p^0}\right)^2 \left(\frac{\partial p^0}{\partial t}\right)^2 \tag{8}$$

so that the field equation (6) now becomes:

$$\nabla^2 \phi = \frac{4\pi G}{c^2}\rho W_0^2 + \frac{1}{p^0}\frac{\partial^2 p^0}{\partial t^2} - \left(\frac{1}{p^0}\right)^2\left(\frac{\partial p^0}{\partial t}\right)^2 \tag{9}$$

This is strikingly similar to (VI-5). To make the identification complete, put $p^0 = mc$ and $W^0 = c$, and convert the mass factors to mass density:

$$\nabla^2 \phi = \frac{4\pi G}{c^2} \rho W_0^2 + \frac{1}{\rho}\frac{\partial^2 \rho}{\partial t^2} - \left(\frac{1}{\rho}\right)^2 \left(\frac{\partial \rho}{\partial t}\right)^2 \qquad (10)$$

Furthermore, this recovers the 2nd and 3rd terms on the RHS of (VI-5) if Jim's original substitution is used in those terms to set ratios of $\phi/c^2 \to 1$. Missing from (10) relative to Jim's (VI-5) are the quadratic time derivatives of the field. Neither of those terms are important to Jim's Mach effect, so it appears they are dispensible parts of his theory. I like this simple derivation here because it reproduces the essential parts of (VI-5) without any assumptions or without the uncertainty of where to substitute for c^2.

Indeed, in hindsight we can recover equation (10) from Jim's equation (VI-4) if the 3rd ansatz is dropped. Taking straightforward derivatives of the first term on the LHS of (VI-4) yields exactly equation (10). This seems to imply that the 3rd ansatz is outside GR, and it accounts for the extra time derivative terms in Jim's derivation.

Jim feels the extra time derivatives are important, especially the piece of the d'Alembertian, to make the Mach equation look relativisitic. I am not so bothered by this because we know the Newtonian limit of GR is not covariant, and we know the linear limits of GR are not covariant. We can still get relativistic effects to a certain order, but without a covariant equation.

Mach Effect Discussion with J.F. Woodward

Editor's Note: Professor Woodward's Estes Park Proceedings paper is not licensed for Open Access. It is only available from the Space Studies Institute web site, as part of the full James F. Woodward commemorative volume. The discussion from Professor Woodward's session is captured here.

During Professor Woodward's session, he shows a hand-drawn graph of force versus voltage V, showing a V^4 dependence. There are some strange looking "errors bars" in the form of ovals drawn on the graph.....

Hathaway: Why are the error bars larger for larger thrust levels?

Woodward: Those are not error bars in the usual sense (like RMS values); the width of those ovals represent the spread of voltages I used for that data point – there were several runs taken for each point and a corresponding small change in voltage for each run.

Williams: What would it take to get to Newtons of thrust, is the scaling practical?

Woodward: Yes the scaling is practical, but I wasn't thinking of going straight to heavy lift. We should increase the frequency to MHz rather than kHz, that would do the trick. For thousands of Newtons we would need GHz frequency. Right now we have $\sim 1\mu N$ of force say, that is at 35 kHz. We believe the force is dependent on frequency f between f^2 and f^3. It's difficult to say exactly how the force scales with frequency. There are a lot of parameters in our equations which change with frequency, so as the device heats up the natural frequency changes and all those parameters change too. If you go through all these terms and extract the frequency you get a force that appears to go like f^6 – all you have to do is go up in frequency to GHz and I could move my house ...

...audience laughter ...

Woodward: It's not so simple I'm afraid. But running at higher frequency is clearly a desirable thing to do. Let's assume the force goes like f^2. So if we increase the frequency from 35 kHz to 35 MHz, with everything else remaining the same, then the new force is $F = F_0(35 \times 10^6/35 \times 10^3)^2 \rightarrow F = F_0 \times 10^6$ then the new force would be $F \sim 1N$. If instead I used 3.5 GHz then $F = 10KN$. Right now my equipment can only manage the kHz type frequencies, I don't know if these devices have resonances with that high a frequency. Whether the devices will end up looking like PZT stacks or some other solid state device or like an EM-drive is not obvious to me yet. I would not be surprised if they turn out to be solid state devices like the ones we are running. Cavities have a lot of extra baggage that is not present in solid state devices. The optimum running conditions for any device turns out to be ~ 1 GHz. If you get too much above that frequency then the ionic response of the material cannot oscillate fast enough. You want the heavy ions moving not the electrons, they would decrease the magnitude of the effect by 3 orders of magnitude or so.

Williams: Are these devices only space drives in a sense they only work in space or are they capable of heavy lift on the ground?

Woodward: At the moment we are talking about space drives because of the low thrust levels. But eventually these drives could be developed so they can have heavy lift capability.

Williams: How big do you think these heavy lift drives will be?

Woodward: I expect the size to be similar to what they are now, maybe a little smaller – but we would be using arrays of devices possibly with redundancy and plug – in replacement capability. We might cycle them on and off to reduce fatigue and failure and make them last longer. On a space mission we would carry 20% more devices than needed in an array to compensate for failures and have added redundancy array systems in place. Trying to scale any device to give a force bigger by a factor or 5 (or perhaps 10) is a fraught with difficulty and a bunch of unknowns. There is a lot of hard work ahead. We have only just started on a journey which could lead to heavy lift eventually. This is not going to happen overnight. At least we now know the physics which points us in the right direction and suggests what we should try next. We have a clear path forward, and I think it is fair to say that as short as 6 months to a year ago that was arguably not the case.

Mach Effect Discussion with J.A. Rodal

Editor's Note: Dr. Rodal prepared a full monograph instead of a proceedings paper, but it is not licensed for Open Access. That monograph is available from the Space Studies Institute web site, included as an attachment to the full James F. Woodward commemorative volume. The discussion from Dr. Rodal's session is captured here.

Dr. Rodal presented a comparison to measurements of his calculations of expected thrust from a Mach effect device. He started with a history of the precursors to the Mach effect device, going back to the discovery of piezoelectricity in 1880 and the subsequent development of the Langevin ultrasonic transducer. The original Langevin transducer, and the piezoelectric devices used today, involve stacks of piezoelectric material because the material is quite brittle.

Fearn: Regarding the eletrostrictive parameter of the lead zirconate titanate PZT (Steiner and Martins SM–111, a modified form of PZT–4 or Navy Type I), there are very few references that have the value of the electrostrictive parameter of PZT-4 in them. Your equation shows how you can experimentally determine the value for electrostriction for a given stack at a certain temperature and frequency.

Rodal: Yes, I only found 3 references that had enough data on experimentally measured values of electrostriction for PZT formulations to ascertain an estimate of the electrostrictive parameter of hard-doped PZT.

Meholic: Does the natural frequency change with temperature, so as you run the device would it change natural frequency as it heats up?

Rodal: Yes – the natural frequency will decrease with higher temperature (since the stiffness decreases with temperature) and will change with thermal, electrical, and stress-strain history.

Hansen: How do you know how much to tighten the bolts?

Rodal: Excellent question. It should not be on a torque basis, it should be on the stress needed, informed by an engineering analysis.

Meholic: After the devices have been run for a while, and shear cracks have developed from the compression, do you need to re-tighten the bolts after the device cools to prepare it for the next run?

Rodal: A better way to do this is to run impedance spectroscopy, as I will show later.

There is discussion that because the resonances are narrow in a Mach effect device, one would want to employ automatic frequency control...

Meholic: What would you use as input to control the frequency?

Rodal: We will talk about that later.

Buldrini: I am unconvinced that the bracket [supporting the test device] has no effect on the frequency, and that it becomes effectively decoupled from the stack.

Rodal: I took a good look at that. Either by luck or as a result of trial and error, the brackets in use are thin enough so that the stack behaves as a free-free resonant spring with lumped masses attached at its ends. The support is not stiff enough (compared with the stiffness of the stack) to act as a stiff mechanical clamp. The bracket is able to flex and accommodate the natural frequencies of a free-free stack. We actually tested this, we used a piece of very thin aluminum as a bracket so thin it was easy to bend by hand and Heidi was worried it would not support the weight of the stack. Heidi ran one PZT stack with brass tail mass and aluminum head mass on Keith Wanser's SR-780 impedance analyzer with the ~ 0.72 mm thick (2.7 g) aluminum bracket and ten separate runs of the regular ~ 3.21 mm thick (6.8 g) aluminum bracket and all tests gave the same impedance spectrum. So we are quite sure that the bracket is effectively decoupling the device from the balance beam and is not significantly influencing the natural frequency of a free-free stack.

Broyles: What were the bolts made of that hold the stack together?

Fearn: There are 12 stainless steel bolts. Six 4:40 cap screws attach the brass to the mount bracket and six 2:56 cap screws run through the aluminum end cap on the outside of the PZT stack and enter the threaded brass mass. These hold the stack in place and have heat shrink around them for electrical insulation.

Broyles: I presume your problem with thermal expansion will be with the PZT stack itself?

Rodal: The main problem with thermal expansion is that the natural frequency is shifting, and the Mach effect force is extremely sensitive to it. The second main problem is damage. The change in temperature will alter the electrical properties.

Meholic: It appears the only cooling, at the moment, is at the ends of the stack, by the brass mass and the aluminum end cap.

Rodal: The cooling is mainly through the ends. But ideally heat would be transported out the side surfaces as well.

Mach Effect Group Discussion

Editor's Note: The preceding sessions on various aspects of Mach effects produced such zeal for further Mach-effect consensus building that the Block 4 scheduled session was cancelled, and Block 4 was devoted instead to this follow-on discussion of Mach effects.

Bushnell: I just have a couple of observations. A few of us came here to really carve out the Mach effects and the EM drive details. We have a lot of people asking us about this stuff. It is incredibly important to get this right. First of all the retro-causation theory that we heard of, this is one of many interpretations of quantum mechanics. There is no canonical interpretation of quantum mechanics, there is no agreement on the correct version of quantum mechanics.

We heard this morning what I think is a superb presentation by George Hathaway. I asked the JSC (Johnson Space Center) people (namely Eagleworks), they have a chart we could go over, they had an experimental chart with some of the experimental artifacts and issues on their particular thing, (EM drive) I didn't see a similar chart, maybe George has one for Jim's (James Woodward's) Mach effect work.

So I asked Jim, have you been through all this, because you say the people at JSC have been beat up by the JASONS (an independent group of elite scientists which advises the United States government on matters of science and technology, mostly of a sensitive nature) and some external committees that JSC brought in, to look at all this. Jim replied he had not had time to go through them all. So the current state of play seems to be, that the amounts of force from the Mach effects are still fairly small, that all of the experimental artifacts and issues have not by any means been addressed and unless and until they are, and the importance of this is so massive, that I think we deserve to (for the scientific people as well as society) do our due objective diligence on this, just as well as we can, before we decide what is right and what's wrong, what's real and what's not OK.

The issue of the Mach effect. You know I've been an engineer I'm not a physicist, so I go read stuff. The vogues in physics are not particularly kind to the Mach business. So along with the retro causation, there's the whole issue of the viability of the Mach approach. So there's a lot of issues here that I think need to be discussed going forward and the people in this room, a great many of them, are probably the best in the country to do that.

This is why I prevailed and fussed at the organizers. They came to me and they said no no we got a schedule, we got to move forward and do this and so forth and I looked at them and I said this is supposed to be a technical meeting, we're not here to salute schedules as far as I am concerned. We're here to sort out what's really going on with all this stuff. Thank you for listening.

Fearn: Dennis Bushnell is correct, this is a technical meeting and a workshop. The schedule is not as important as the discussion we are here to promote. We were supposed to leave time after the talk (and during) for feedback and discussion. So in order to allow for Q&A we are going to postpone the rest of todays talks and continue right now with a Q&A session for both José, myself and earlier talks.

Woodward: Can I start off with a question to George? I think I may have missed something that was said about a replication [of the Mach effect] using a device that I sent you. I heard later on, near the end of your talk, you said you saw a $0.2\mu N$ thrust which is about the same thing that Nembo got with a similar device that I sent him. Did you see the transients as well?

Hathaway: Yes, I called them "pips" in my talk, they were not as pronounced as yours but they were in the right direction.

Woodward: Were you using the picoscope software? (a small device which hooks up to a computer via USB and generates an oscilloscope like screen on the computer monitor)

Hathaway: No, I was using a standard oscilloscope, a real analog type scope. You know, one of those old things with a funny looking small green screen...

Woodward: I remember storage oscilloscopes where you had to pull a handle down...

...laughter...

Hathaway: Well not quite that old, I used an analog scope and some digital ones too, but I like the analog scopes when there are fast transients. If I want to see a fast transient I throw the digital scope out of the window. But I did see the "pips", as I called them, the leading edge and the trailing edge on the thrust trace.

Woodward: Okay, and theoretically the reaction mass you have was the 5/8" brass mass and Heidi gave you the new 3/4" brass mass to use?

Hathaway: Yes, I presume take off the old brass mass and substitute this new one and everything else stays the same?

Woodward: Yes that's right.

Buldrini: I would recommend you let the device sit in the vacuum chamber for some time before making a new measurement because the PZT (lead zirconium titanate) tends to absorb water vapor which reduces the Mach effect force.

Hathaway: Yes and I'll make sure I re-torque the bolts to 4 inch pounds as Jim recommends.

Woodward: That's 4 inch pounds for the 2:56 bolts and 2.5 inch pounds for the 4:40 bolts. I just wanted to make sure I didn't miss anything important this morning.

Hathaway: What you did miss was my many concerns about prosaic effects and one of them was something you brought up. You mentioned the Lorentz air effect where there is a possibility that you can give enough momentum to enough air molecules that they collectively produce a force that is measurable. That was one of the issues I forgot to mention.

Woodward: The way that is taken care of is by simply running the device at different pressures because the Mach effect does not change with pressure. The Lorentz air force however, would be greater at higher pressure.

Rodal: It is extremely important to characterize what it is that you are going to measure before you run the test. Otherwise one is flying blind. That piezoelectric materials have history dependent properties has been known for 100 years. The PZT may lose its poling. Every time you conduct a test you are going through cycles where you may cause damage to PZT, which is a very brittle sintered ceramic, and you need to characterize its state of damage. I would run an impedance spectrum and I would look for the natural frequencies. It should be that the natural frequencies are staying the same. If those frequencies are getting lower every time, you know that you have damage. The mechanical natural frequency depends on the square root of the stiffness divided by the mass. The mass is not going to increase with cyclic life. The only reason the natural resonances would decrease is due to damage; it is caused by the stiffness going down with time. This happens because there are cracks emanating from voids in the sintered ceramic PZT and they cause a decrease in the resonant frequency. So when you run the device again and again and see the force decreasing every time it may be due to damage. These sintered ceramic materials are sensitive to fatigue damage. Conducting tests on a stack without immediately prior conducting impedance testing is like a physician conducting a stress test on a patient without prior measuring her blood pressure, pulse and other physical tests. Also this goes for any statistical analysis that plots variability without taking into account the initial state of damage. It is not right really to compare the results of healthy new stacks to old stacks. For example, let's suppose that that Jim measures a new, young vital stack and someone else [George or Nembo] measures an old decrepit stack, clearly the young one without all the damage is going to do much better.

Hathaway: To add to that comment, something that was glossed over, there is an accelerometer in the stack, that is also a good, subtle way, of looking at how the stack maybe deteriorating. The comparison between the input waveform and the waveform that is produced by that accelerometer at the beginning of your series of experiments versus say the the middle or the end of your data set because that accelerometer itself can change characteristics because it is inside the stack, it's the same material as the stack.

Woodward: Yes, the accelerometer is just a thinner PZT disc 0.3mm thick, rather than 1 or 2mm thick for the rest of the discs.

Hathaway: Right, so they are of the same material only thinner, so they are even more subject to change than the thicker discs.

Meholic: How do you get the signal out of the accelerometer?

Fearn: There are wires attached to the electrodes on either side of the accelerometer. The accelerometer is the only disc in the stack not receiving power. We take out the piezoelectric voltage, caused by contraction in the stack, and use that as an indication of stack acceleration.

Meholic: So you have electrodes inside the stack of PZT discs, made of brass, that must change the whole integrity of the stack. Jose drew the stack, in his model, as if it was a spring, now you have this different material in the stack which must change the properties?

Rodal: Right, not only do you have damage due to initial voids in the PZT that coalesce into cracks, but you have to consider the unequal stiffness of the segments in the stack, which translate into interface stresses: the PZT, the epoxy adhesive and the electrodes have different stiffness. Therefore you also have, due to cyclic fatigue, damage at the PZT/adhesive/electrode interfaces. You have a spring which does not have equal stiffness in each segment of the stack. That segmented nature of the stack makes the system more complex.

Meholic: So it sounds like, what you have been saying, that these stacks have a very limited life span, because the material breaks down that is key to producing the Mach effect.

Woodward: In test devices that is true, but when these things go into production, and you know the thing is working the way you think it is, you'll be able to cool them and keep them in running order much longer than it is possible when it's just an isolated device in a vacuum chamber with no real heat sink attached to it.

Meholic: But, independent of the cooling, you have high cycle fatigue that was mentioned to worry about.

Rodal: Wait, remember we have 100 years of transducer knowledge for sonar, and there is a lot of data on this. With a commercial device you can go through billions and billions of cycles. But look at what we have here. Here we have a stack where every piezoelectric disc has been bonded with epoxy adhesive by hand, with an unfilled epoxy mixed and cured at low temperature (120 degree Centigrade) for 1 hour, hence having a low Tg (Glass transition temperature).

Meholic: So my basic question was assuming those stacks show a real effect, then you can scale these up? Is the science of piezoelectric materials sufficient to allow you to have a long life thruster?

Woodward: Yes

Rodal: Yes, but you have to be skeptical like Dennis was saying. You cannot assume that every problem is going to automatically take care of itself. We know this fatigue is a well known problem, and it has been tackled before with success. How has it been dealt with? You can either co-sinter the electrodes directly to the piezoelectric ceramic so you don't have the bonding problem or use an adhesive with fillers (that raise the adhesive strength and stiffness as well as thermal conductivity and reduce the thermal expansion). Another approach is to have a single crystal so that you avoid the grain boundaries and voids that promote crack damage. Another concern has to do with the preload. At the moment, the preload is kind of empirical isn't it. [To Woodward] You have not run a finite element analysis to look at fracture mechanics parameters like KIc and KIIc (stress intensity factors, in crack opening mode and in-plane shear mode, respectively.) and determined what kind of preload they need to have. This kind of analysis can be done and has been done in the aerospace industry. If you do that, then the device can be taken to a very high number of fatigue cycles and if run below the fatigue endurance limit one can talk about infinite life. But this requires using a high level of technology to do the analysis and construct the stacks. [Rodal refers to self healing materials]

Meholic: Okay. Here's another weird question for you, Jim. Assume this is based on my primordial understanding of what you guys are doing. So this stack changes its energy content and so essentially changes its mass and the universe reacts accordingly. In the same vein, if it's changing its energy and this changing its mass, does the Earth react on it accordingly to give it a vertical component of displacement?

Woodward: Yes very slightly

Meholic: Why is it very slightly vertical and more horizontally?

Woodward: Because the Earth's gravitational field at the surface is only GM/R for potential energy. The universal potential is $\phi = c^2$ for flat space, which is a lot bigger.

Fearn: The property of inertia is due mainly to the distant matter in the universe. That is what Einstein believed, as did John Wheeler.

Hathaway: Can you guys summarize for me, I think we have at least 3 different methods of arriving at Jim's mass change formula for the Mach effect. We have Jim's, Heidi's, and José's is that a correct evaluation?

Woodward: There's a fourth, that is Lance [Williams] has another way, from linearized general relativity.

Fearn: Yes that's about right. I used Hoyle and Narlikar gravitation which is a fully Machian theory. It is fully covariant and reduces to Einstein's general relativity in the limit of a smooth fluid, in the rest frame of the fluid. I found Jim's mass change equation in the Hoyle Narlikar field equation for a smooth fluid mass density, together with some extra spatially changing mass terms.

Mathes: Heidi, can you write an executive summary, two pages that explain the Hoyle Narlikar theory simply, so a non specialist can understand the derivation?

Fearn: I have some typed up notes from reading the Hoyle Narlikar papers and book, but it's about 60 pages...

Meholic: That's brief for Heidi...

...laughter...

Rodal: There is a difference, because I was looking very skeptically at this at the beginning and I was able to reconcile this looking at kinetic energy, where I can see a clear path to get through it. I am not happy in the derivation where you put $E = mc^2$ and translate into a changing mass. Einstein's first $E = mc^2$ proof was shown to be a circular argument. Einstein himself was so dissatisfied by it that he had a number of improved derivations throughout his life. I can only see it when I consider kinetic energy, then I can see it step by step. I think it is very interesting that the experiment is showing a force in the same direction as the calculation. I can also calculate an optimal tail (brass) mass. This could be a coincidence, but I find it very interesting at this point. But I am not there yet because to match results I need to use a single coupling factor of unknown origin. I think that there are a lot of nonlinear terms that are being dropped and those need to be examined in more detail.

Hathaway: Is there any overlap between the derivations of Jim, Heidi and Lance or are the theories totally independent?

Woodward: They are three variants on making one or two assumptions. One assumption is that the rest mass is not a constant, the rest mass can change with time. The other is that inertia is a gravitational effect and those are the only two assumptions you need. In effect the three derivations you are talking about are three ways of getting to the same answer. Simply a choice of approximation.

Williams: But they come out somewhat different. I get something a bit different, but in essential aspects they are the same.

Hathaway: Looking at it from the view point of a physics journal editor, these three papers come in, and they are all essentially claiming the same bottom line. I'm thinking logically are they self consistent, are they saying the same thing from different viewpoints and so you should put them all together or state the individual assumptions that are similar or different than the other two.

Woodward: Let me give you an example, Lance versus mine. What Lance does, he derives the result from the geodesic equation and in order to get the result he makes an assumption about the time dependence of the field quantities in favor of the time dependence of the source quantities.

Williams: I just used standard linear general relativity, like in gravitational waves or the Newtonian limit.

Woodward: You can read Lance's version, it's in the prep kit he sent out and you'll see what I mean. You'll find there that he has a comment about how his version differs from mine because I insisted on keeping the time dependent quantities as field quantities so I would get the d'Alembertian of the potential is equal to the sources, he gets the Laplacian of the potential is equal to the sources. But the time dependent sources are basically the same thing. So it's a matter of his choice of approximation versus my choice. It's not a matter of fundamental elementary physics. He's not talking about anything fundamentally different than me. Would you say that's fair Lance?

Williams: Yes that's sounds fair.

Hudson: It sounds like some high level comparison between differences and assumptions

Woodward: No, it's just the difference between a theoretician and an experimentalist...

...*laughter*...

Hudson: That's okay but I still think it might be useful.

Rodal: Yes it would be nice to know what the assumptions are. You just said there is a difference in the assumptions being made, I would like to know.

Woodward: You can read Lance's paper too and you'll see what I mean.

Williams: Also, there is a quick study that summarizes Jim's derivation, and it has all the assumptions that Jim made, straight out of the stargates book [Referring to "Making starships and stargates", Springer 2012, by J. F. Woodward]. My version does not have any of the assumptions that Jim has, it's just mechanical linear general relativity.

Kelly: You said there were two assumptions. One was that inertia is a gravitational effect. Isn't there some kind of experimental test to show if the Mach principle is true? Since the universe is expanding wouldn't inertia (from Mach's principle) be different in the past from what it is now? Isn't there some distant object you could look at to show if inertia has changed?

Woodward: As it turns out Wes, that's not true in general relativistic cosmology. Because you are choosing one cosmological model, it's one that is spatially flat and borderline between open and closed universes and that has the odd property that the value of Ω, which is the measure of spatial flatness ($\Omega = 1$ in this case) is the same in all epochs. So all of this is automatically compensated for. Inertia in early epochs and inertia in much later epochs than ours, will be the same if $\Omega = 1$. You can do experiments to see if $\Omega = 1$ or not.

Kelly: The whole idea of inertia seems a bit slippery to me

Woodward: Yes I agree, it was slippery when I first came across it too.

Hathaway: Are there any other experiments other than force experiments that could validate Jim's theory? Anything that does not involve measuring small thrusts?

Woodward: Yes.... Big thrusts!!
...laughter...

Woodward: No seriously, what we really want to do now is find a way of scaling the thrust, with the merger resources we have, as quickly as possible. Indeed, I've already talked with Jose about this. I'm going to be ordering a bunch of Steiner-Martins crystals sometime in the next week or two and perhaps you could join in the conversation about what would be the ideal configuration for those....

Hudson: What bothered me about your talk [addressing Hathaway] this morning, was all these tiny effects you were trying to mitigate by proper experimental design. But if you put a rocket engine on a test stand you won't pay attention to the rotation of the earth or the angle of the sun, these are obviated completely by the scale of the thrust. A demonstration of an ambiguously larger thrust that would eliminate the concern about many of these small experimental error sources, that you described, would go a long long way to making the Mach effect look real to the public. This is why I have always encouraged more thrust. I know a lot of people in this room view rockets as the best way of getting into space, and I agree with this, but I think the problem of energy dissipation in these Mach devices, perhaps the problem of fabrication might actually be solved by the technologies we use in the liquid rocket industry. I would like to get the few of us with liquid rocket experience sitting down together with the physicists to engineer a new device that could incorporate some of these new technologies. We can do things today that were impossible ten years ago. Things like diffusion bonding, small channel laminar flow for cooling...

Hansen: These new technologies will change the resonant frequency of the device. There are more ways of cooling than with rocket technologies. A rocket engine runs at several thousand Kelvin, a PZT stack won't get anywhere near those temperatures, it would no longer function long before those temperatures are reached. So there are other ways to deal with it. My main job is working with radar. They are actively cooled, some are air cooled. We also look at duty cycle, which is the amount of time you transmit versus the time you don't. Some of that is based on heat, because you can't heat up the radar components more than a certain amount. We have to calculate how much heat we have to take out before we do any cooling because we don't want to spend too much money on a solution.

Turner: I would like to follow through with Gary's idea of where we could go with this. You mentioned to Jim that there were things you could do to increase the thrust level substantially. So if I look at tens of micro Newton of thrust requiring a 100W of power and assuming power is 10W per Kilogram I don't get a very big acceleration and I haven't put a payload in yet. So what are the thoughts on how to scale this to something that is practical?

...discussion about packaging of multiple wafers onto a sheet of silicon and discussion of possible arrays of devices that have not been tested yet...

Turner: But there was talk about even the current devices producing more thrust, how would that work?

Rodal: Yes, the first order of business is to improve the device we have working now. Correct me if I am wrong Jim, but you are using guitar amplifiers and set the frequency by hand with no automated control?

Woodward: Yes that's right.

Rodal: To me, the first order of business is to have automatic frequency control. This would allow you to track the resonant frequency with a very small bandwidth. The bandwidth you need to be at resonance has to be considerably smaller than $1/(2Q)$ where Q is the mechanical quality factor, with the value of Q about 1000. The people that are working on the EM drive, at much higher Q, are already using automatic frequency control. This may get you about 10 to 50 times more force. We want 1000 times more force, so how do we get it? That is where I see the nonlinear terms that were dropped out of the equation coming into play. The terms that have been dropped may play a very important role especially in the damping term (the speed term). Most people neglect damping altogether and just say it is very small. But assuming zero damping means an infinite amplitude response, which is absurd. There must be damping because of the second law of thermodynamics. What does neglecting these nonlinear terms do? Well it may result in a Mach effect force which is much smaller than it could be, not by 10 times, by 100 times, possibly by orders of magnitude. How could it be a million times? By parametric amplification and self excitation, considering nonlinear dynamics. Many types of rocket engines have exhibited self-excited vibration. They are called POGO vibrations, the Saturn V had a number of bad episodes due to POGO self excited vibrations. All this knowledge of self-excitation has not yet been applied [to the Mach effect drive]. I would like to explore how a Mach effect force can be amplified by nonlinear, self-excitation, and that's it.

Hudson: Do you have an idea how to take advantage of this nonlinear effect?

Rodal: Yes I do. The trouble is, to take into account all the nonlinear terms that have been dropped out, will result in an exponentially larger number of terms, that would demand very large computational resources in terms of memory and computing time. At the moment I am calculating $20 + 269$ terms, where each term involves a large number of terms, this is no problem at the moment. However, to take into account the neglected nonlinear terms would result in an exponentially larger number of terms. Is anyone here using Mathematica? If you try to use Mathematica to analytically solve expressions with a larger number of terms you know what a hard time Mathematica has to deal with a large number of terms.

Mathes: We can rent time on a cluster or supercomputer to deal with whatever you can throw at us. We can buy a supercomputer off the shelf at Amazon these days. It's a big problem but one that money can solve.

Hathaway: How far will the nonlinear code go to suggesting new experimental approaches for increasing the Mach force?

Rodal: When you run your Mach effect experiment, what are you doing with the frequency? Can you assure me that you are right on resonance? And as the resonance frequency changes due to thermal and nonlinear effects are you changing the frequency to match?

Hathaway: I'm using the same manual method as Jim is using.

Rodal: Well that's no good.

Hathaway: Absolutely, I agree... Automatic frequency control is clearly the first thing to do, but parametric amplification can come in a number of different forms, and sometimes it comes with different materials.

Materials can change in a way you can exploit to get your parametric amplification. Sometimes its a DC electrical signal superimposed on an AC signal. Would throwing a large amount of money at the computer problem allow the experimentalist to extract from that a way of conducting the experiment to achieve more force?

Rodal: Are you familiar with the Van der Pol equation, that is a good example. We need to consider the additional terms as nonlinear additions to the equation. Once they are taken into account we can say what the experimentalist should look out for.

Fearn – note added in proof: In dynamics, the Van der Pol equation represents an oscillator with a nonlinear damping term. It evolves according to the equation, $\ddot{x} - \mu(1-x^2)\dot{x} + x = 0$, where x is position and μ is a scalar indicating the strength of the nonlinear damping. This equation has a simple representation in terms of a small circuit shown in the Figure below.

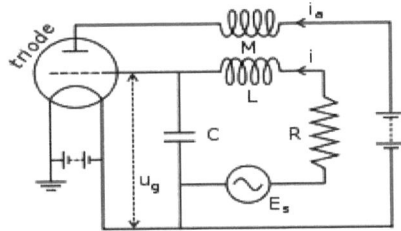

Figure 1: This is a simple Van der Pol Oscillator using a Triode

Hathaway: In the case of the Van der Pol equation, it could be interpreted as a little oscillator circuit and you could see the amplification going up like crazy.

Rodal: Exactly, if you can, and I don't know if you can, but if you can get self excitation in the mass fluctuation then it will proceed like in the Van der Pol equation and the force will increase dramatically.

Hathaway: In that case, you are taking known off the shelf components, putting them together in a particular way, and getting an unusual result...

Rodal: Yes that's right.

Hathaway: And the "unusual result" is perfectly calculable through the Van der Pol equation.

Rodal: In that case you know the exact equation, but in the Mach effect case we don't have the exact equation because a large number of nonlinear terms were dropped.

Hathaway: In these other systems, if you dope the piezoelectric material, like these PZT discs, you might enable parametric amplification that way.

Rodal: This is an excellent point, I completely agree with you. As far as I know, Jim has tried all kinds of devices before the piezoelectric one, you tried capacitors as well right?

Woodward: Yes...

Hathaway: Certainly, the first step is to introduce automatic frequency control, but perhaps you can focus in on exactly which nonlinear terms could produce an approach to an engineering solution for parametric amplification.

Rodal: I can certainly analyze this problem and perhaps come up with something in a few weeks. I have solved problems like this at MIT and also in my professional life. I worked for a company that made very large industrial machines which had all kinds of self excitation problems. It was a big headache for them that we were able to solve and gain a big advantage on our competition. The machines had polymer components that had nonlinear frequency-and-temperature-dependent damping characteristics that resulted in self excitation which increased the magnitude of vibration exponentially under certain operating conditions. Normally you need to eliminate mechanical self excitation problems. This is the first time, I have a problem, where I need to enhance the self excitation!

Hathaway: José mentions a good point, you need to be able to throttle this parametric amplification or you might "blow up" your device.

Rodal: Jim is presently running chirps instead of steady-state resonance. I don't want to self excite to the point that I destroy the stack. These self excitations grow exponentially, they follow a curve and you can stop it at a certain point. To stop them, all you have to do is to cut off the voltage to the stack.

Hathaway: The chirp and voltage cutoff would be part of the autotune system.

Rodal: Yes, as soon as you stop chasing the resonant frequency the self excitation will fall off. In industry, sometimes these big machines with rotational parts start vibrating strongly and the operator gets scared and slowly reduces the speed just a bit, which makes matters worse due to the nonlinear nature of the vibration backbone curve - which is of course the worst thing you can do - it usually makes the vibration worse - you should rapidly increase the speed to reduce the self excitation. But naturally, people are afraid to do this.

Tajmar: I would like to throw in a comment. You spoke a lot of electrostriction, there is also magnetostriction. This introduces a totally different material, there may be no cracks in this like in the piezoelectric you are using.

Rodal: So Martin just reminded me of something here. The frequency dependence of the force. If we look at the equations we see a ω^6 and an ω^{10}, you think wow, our problems are solved, just go to higher frequency. But it is not so simple. You have damping and the equations are complex. Several parameters depend on frequency so really the frequency dependence for the force is more like $\omega^{5/2}$ or somewhere between the second and third power. This is what my solution shows when one calculates specific examples.

Fearn: Experimentally, we can show the increase of force cannot exceed ω^2 or ω^3 at most. The frequency range, that we have data for, is between 29KHz and 40 KHz and so we would have noticed a huge increase in force if the frequency dependence was something like ω^6, and we haven't seen that.

Rodal: So we are confident that the experiments show that the force should go up with something like the square of the frequency, so we should go to higher natural frequency devices. The quickest way to do that is to make shorter stacks.

Tajmar: Then we have a much higher dissipation rate, perhaps a higher voltage would also work.

Rodal: The voltage increase would work, the force goes like V^4 but that is not attractive from the point of view of power consumption... If you go to one or two discs there is another problem with the frequency. These discs are expanding not only in the longitudinal direction (the thickness direction of the PZT discs) but also in the perpendicular direction, the radial direction of the stack. The radial resonances are due to shear deformation and result in similar or lower natural frequencies than the longitudinal resonances for the very short stacks, so you may excite the wrong mode. For the Mach effect you need to excite the longitudinal mode. We haven't tried yet going from 8 discs to 4 discs which might be better. There are also many other piezoelectric materials that we could test, other than PZT, doped materials. A lot of things could be done if we have enough people and divide up the work.

Hathaway: Jim did you ever try using quartz crystals in any of these stacks?

Woodward: No. I inherited a collection of PZT 19mm discs from a nearby industry. They had a fabrication facility just to the south of campus. They were closing down and phoned up the university asking if anyone could use a batch of PZT discs - so I took them.

Hathaway: I'm wondering since the quartz crystals have a much higher mechanical quality factor than the PZT stacks. Quartz has a lower piezoelectric action than the PZT but it maybe that the higher quality factor trumps it.

Rodal: But we have historical experience from the piezoelectric transducers for Sonar. They started with quartz 100 years ago and they moved to PZT more than 60 years ago. However, for the transducer you also want higher displacement.

Hathaway: That's true, but we are not looking for high displacement for the Mach effect.

Rodal: That's right, it is not necessarily the case that what makes the best transducer will produce the largest Mach force.

Hudson: Have you looked at some of the latest sonar transducers? I'm not sure that the Navy is still using PZT's any more.

Woodward: They may well be using a PZT-PMN composite material that came out about 5 years ago. It used to be that you could buy pure PMN by itself and now you can't.

Hathaway: You have to make it yourself, like we did.

Woodward: Yes, but you can purchase PZT-PMN composite materials, it's just a lot more expensive than the Steiner Martins (SM-111) material I've been using.

Hathaway: Another question for Jim: what is the downside of making a boat, instead of your Faraday cage, and filling the boat with a dielectric fluid to put your stack into, for cooling purposes?

Woodward: Once you are satisfied that the stack is really producing thrust, and there are probably half a dozen people in this room that figure this is really a good bet - at that point talking about any kind of cooling process that doesn't mechanically disrupt the function of the device will do. Since you don't need to worry about Lorentz forces in the fluid that might be making something appear as a force of non Mach origin.

Hathaway: The next question is, can Jose's equations handle a fluid bath of a material of a specific density? So I'm suggesting we immerse the entire moving body in a bath of dielectric fluid.

Rodal: As a skeptical experimenter, if I take a powerful probe (like a scanning electron microscope) and take a close look at this material stack, I will see a lot of little voids, and eventually cracks that may be interconnected. Now I'm going to be putting this stack into a fluid and I know that when I do this, when the stack expands a crack will open and the liquid will flow inside. Now when the stack goes into compression I have that fluid there, that is going to expand the crack and cause damage at the crack tip, further expanding the crack. So the result is damage. Damage that results in lower natural frequency, lower stiffness and eventually failure. I need some way of keeping the fluid out of the cracks. You also need a low viscosity fluid because of the damping.

Hathaway: Yes, I was thinking of a low viscosity fluid with a high thermal conductivity. Can this be fit into your computational model? Can you show that it would not damp the stack too much?

Rodal: Yes it can be worked out. You have a nonslip boundary condition between the cylinder and the fluid. I don't think it would be a major problem to model. You would have to use a non-permeable material interface. Epoxy is not acceptable. Epoxy, like most polymers, is quite permeable to fluids.

Woodward: Instead of using a liquid you could use Helium gas at room temperature. Helium has fabulous thermal properties.

Woodward: The other thing is, given my slight knowledge of the PZT business, they have very elaborate encapsulation techniques for commercial PZT's. If you set a non-coated PZT disc out in a humid environment, the disc will absorb water vapor and that kills it. We have been keeping our PZT's in sealed containers with desiccant or inside the vacuum chamber to preserve it. When the discs absorb water vapor they do not function very well. I discovered this by accident a few years ago. In some cases you can reverse the damage just by putting the stack into the vacuum chamber and pumping on it for a month. It gets the water vapor out of the cracks.

Hudson: It would be better from the rocket engineer point of view, to have a solid state cooling solution. I wouldn't want to put a fluid on a rocket undergoing high g-forces.

Hathaway: Oh no, I'm talking about for laboratory experimental purposes. My basic question is whether Jose's equations are amenable to adding this kind of fluid bath and noting if it might cause too much damping to be an effective cooling method or not.

Rodal: Yes it is doable to model it.

✔ Fuel Problem	Theory	Existing Physics
Time-Distance Problem	✔ Experiment	✔ New Physics

RF Resonant Cavity Thrusters

- ◆ Paul March will lead a discussion about experimental efforts at Eagleworks
- ◆ Martin Tajmar will lead a discussion about experimental efforts at Dresden
- ◆ John Brandenburg will offer a theoretical explanation for the effect

Issue Summary:

These devices purport to provide propellantless thrust by emitting RF waves inside a shaped cavity. By the shaping of the cavity, a net thrust can result from the momentum transfer of the RF waves with the walls of the cavity. These are alternatively called the electromagnetic drive (EM drive) by Roger Shawyer, and the Cannae drive by Guido Fetta.

Any such effect constitutes new physics, because our current understanding is that momentum conservation should not allow such a thing. It is similar to being able to make a spaceship go by pushing on the inside. If confirmed, such effects would revolutionize physics. A compelling, repeatable experiment is needed because conservation of momentum is a cornerstone of physics. The electromagnetic field has never been found to violate conservation of momentum.

Shawyer has released a technical paper online [1]. His theory is based on a misconception regarding his equation (1), the Lorentz force law. The Lorentz law gives the electromagnetic force on charged particles; it does not describe the dynamics of photons or RF waves, which are uncharged. Shawyer's theory, at least, appears unsatisfactory. At this time we must consider this effect an experiment with no accepted theoretical explanation or theoretical estimate of the magnitude of the effect.

John will pick up that gauntlet and offer a theoretical explanation in terms of his GEM theory, a unified theory of gravity and electromagnetism. He has done much work on it over the years, e.g., [2].

Martin will summarize experimental work done at Dresden, where they have also conducted a wide array of experiments in related areas. See the web site [3] for additional discussion possibilities.

Like at Dresden, Eagleworks has done a wide array of experiments. But Paul will focus at Estes Park on the RF cavity thrust experiments.

A Wikipedia page has appeared on this topic since April, initiated by an IP address in New York. Reputable people have provided inputs, and the experimental efforts of two of our session leaders are described. This page is annotated and well-written, and so is recommended for background review, even though it only includes journalistic sources.

Our objectives in session will be to understand the experiments in a way that allows peers to verify them, and to understand what theoretical modifications are necessary to recover conservation of momentum.

1. http://www.emdrive.com/theorypaper9-4.pdf
2. J. Brandenburg, Int. Jour. Astrophy. Sp. Sci., v2, p24, 2015. doi: 10.11648/j.ijass.s.2014020601.14
3. http://www.tu-dresden.de/ilr/rfs/bpp
4. https://en.wikipedia.org/wiki/RF_resonant_cavity_thruster

Estes Park Advanced Propulsion Workshop	Block 1,2,3	21 September 2016

Experiments with RF Cavity Thrusters

Paul March
Friendswood, Texas, USA

This talk is about experimental tests of advanced, or breakthrough, propulsion systems that I was involved in at NASA Johnson Space Center. As you know, Johnson is devoted to the manned space program. We all share the ultimate goal of manned missions to the stars, by whatever propulsion system is available to us.

1. The Long-Term Vision

Let me start with the goal, and then return to the current state of experiment. At NASA, we have dreams of manned missions to the stars. An artist by the name of Mark Rademaker [1] came up with a fantastic design for an exploration solar-system or poor-man's star-ship, called the IXS Clarke, based on Q-thrusters for propulsion, see Fig. 1.

Editor's Note: Q-thruster stands for "quantum vacuum thruster", a hypothetical propulsion device that would somehow extract energy from the quantum vacuum. The idea is controversial and its physical plausibility undemonstrated.

The Q-thrusters in this concept design are powered by a ~ 2.5 MW nuclear reactor and work via a quantum vacuum based plasma system that Harold "Sonny" White came up with [2,3]. Roger Shawyer was the first to propose such a space drive, which he calls the "EM drive", see his EM-drive web page [4]. That is what we would like to build.

Editor's Note: the quantum vacuum plasma concept is controversial and its plausibility undemonstrated. The EM-drive stands for "electromagnetic" drive. Shawyer's device does not rely on any quantum concept in its design or construction, and its operation depends only on classical electromagnetic effects of RF radiation. It is essentially a microwave cavity resonator. Since any thrust should be impossible due to to electromagnetism alone, the influence of the quantum vacuum is invoked, and so March considers the EM-drive to be a Q-thruster.

These twenty, Q-thruster "engines" on the IXS Clarke concept vehicle are projected to produce about 30,000 N total of thrust, with up to 1,500 N and up to 100 KW RF power per engine.

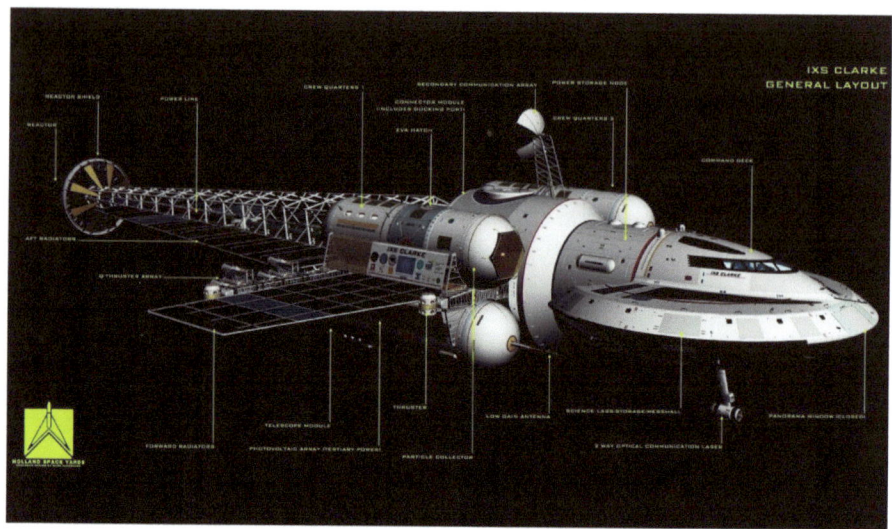

Figure 1: Nuclear powered Q-Thruster Concept Exploration Ship called the IXS Clarke, artists impression [1].

Rodal: How big are those Q-thrusters, how long are they and what diameter?

March: Those Q-thrusters are based on our current design dimensions that are being tested in the lab, at lower power. They are truncated copper cone radio frequency (RF) cavities, with a large diameter of 11 inches, small diameter of 6.5 inches, and 9 inches long. For the concept vehicle, we would pump 100 kW of RF power into these cavities at at frequency of 919 MHz. There are 20 engines, so that will require a 2.0 to 2.5 MW nuclear reactor. This ship would be constructed in space and the engines would be space-drives, meaning that they only work in space. With that in mind, given the total-thrust-to-spaceship-weight ratio of approximately 1.5×10^{-3}, we estimated the starship to weigh about 100 metric tons.

Woodward: Do these cavities have polymer discs inside them? and how do you cool them?

March: Yes the cavities may have thermoplastics discs in them and cooling, well, that's the main trick, isn't it? I think Gary Hudson has an idea how that would work - basically very much like liquid fuel rocket engine cooling. These EM drives are truncated copper cones (called "frustrums"), which basically have the shape of a rocket nozzle. You can cool them like a rocket nozzle. Instead of liquid propellant you are dealing with either a quantum vacuum mechanism [3] or a gravitational field effect [5] to produce the thrust. This of course depends on whose theory you want to use to explain the force.

Figure 2: Copper truncated Cone Cavity, of the type used by Shawyer. Tested at Eagleworks by Paul March 2013. This shows construction of the copper cavity. The large diameter is 11 inches, the small diameter is 6.26 inches and the cavity is 9 inches long. To complete the build 2 type-N RF feeds must be inserted into the cavity. This cavity shows PCB end plates in use. There were 36 # 6-32 nut and bolts used in construction.

Hansen: Would it make a difference to the thrust if you were to use a superconducting material for the cavity rather than the copper?

March: From what we have found to date experimentally, the thrust appears to scale with the input RF power times the quality factor of resonance Q, so if the Q is 100 to 2000 times larger, than what you can obtain for a room temperature version of the device, the thrust should scale up accordingly. Let's see, the level of thrust is currently at tens to hundreds of micro Newton. We hope the next generation of device will improve the thrust to tens of milli-Newtons. Now if you go to a superconductor, the Q jumps from 5×10^4 (depending on the mode you are in) currently to perhaps 5×10^6 (which is a hundred times improved) or to 100×10^6 which is 2000 times bigger. You would expect the thrust to scale with those Q factors. So if we started with a copper cavity device producing 1mN then went to a superconducting device with 2000 times bigger Q, all else being equal, we should have an improved thrust of 2N for that device. So your thrust per power input will go up accordingly. The problem you have, is that now you have to deal with cryogenics. For liquid nitrogen cooled superconductors the temperature is 77 Kelvin. If you have to use a liquid helium then you are down to 4 Kelvin. The liquid helium refrigeration is nasty. So there is the extra weight of the cryogenic equipment you need to carry along. But, there is quite a good increase in efficiency, so you reduce your power level for a given thrust level.

Editor's Note: Q-factor parameterizes the damping of a resonator, and the EM-drive is an RF resonator. Q-factor can be defined as the ratio of energy stored to energy dissipated per cycle; higher Q-factors correspond to lower dissipation.

Rodal: The efficiency goes up with Q, and we have known for a long time that the Q increases with the size of the cavity, so why is your cavity so small? Dr. Luis W. Alvarez (Nobel prize winner) was at Massachusetts Institute of Technology during the war, in the 1940's, developing radar technology for early warning systems. He built a huge 40 foot long cavity using spare WWII equipment. Why is it, that looking into the future, you are using these small cavities, when a bigger one would be so much more efficient?

March: The resonant frequency of the cavity goes with inverse length squared. You can think of an oscillating parcel of air, of mass m, oscillating like a mass on a spring. The longer the length of the cavity the lower the frequency. The thrust increases with the frequency and the Q. That is all part of the optimization process, we would need to do a case study to find the optimal length of cavity and frequency, for a high Q and the maximum possible thrust. That particular unit was based on a 929MHz magnetron, at 200 kW power and running at 88% efficiency. You go with what you've got!. I'm sure if you had enough money you could develop a better more efficient system.

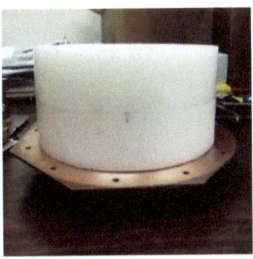

Figure 3: High density Polyethylene (HDPE) discs. Outer disc diameter 6.13 inches and thickness 1.063 inches. Two discs are mounted on the small end of the truncated cone cavity.

Woodward: Assuming that the thrust is real, and that it is being produced inside the cavity by an interaction between the RF field you are injecting and the walls of the cavity, that should give you a means of testing which theory is correct. (Either the Mach effect gravitational interaction theory or Sonny's vacuum plasma theory.) Considering that there are RF photons, that can interact with the skin depth of copper and any polymer disc inside the cavity, that is all you have. So either the force is produced between the interaction between these photons and the walls of the cavity and polymer disc (if there is one) or the force is somehow produced by the electromagnetic field (the photons) independently, without any interaction between the cavity walls or the disc being present. You don't have anywhere near the Schwinger electric field strengths (10^{18} V/m), needed to induce vacuum breakdown and electron-positron pair creation!

Woodward: Do you or Sonny have any plans to convert your copper cavity into a superconductor?

March: I don't have any plans for a superconducting cavity because I don't have the funds, I'm retiring. You would have to ask Sonny what his plans are.

Rodal: What would be the best experimental test to check on which theory is correct?

Woodward: Making a superconducting cavity, because the skin depth of a superconductor is a minute fraction for what it is for a non-superconductor like copper. So a theory depending on the interaction of the photons with the metal material would not predict as large an increase in thrust with

Figure 4: Superconducting Niobium Cavity by Guido Fetta, from the Cannae website 2011 [6].

the change to a superconducting material, even though the Q of the cavity does increase. The thrust may not go up at all.

Fearn - note added in proof: Any induced currents in the wall of the metal would increase for superconducting cavities. This could lead to additional Lorentz-type forces, due to both electric and magnetic fields being present inside the cavity. However, the EM fields would have difficulty penetrating the supercondctor. Overall, there would probably not be as big an increase as expected, with the superconducting material, if the gravitational Mach effect is responsible for the force.

Woodward: Have the Cannae people [6] built a superconducting cavity yet?

March: Guido Fetta built his first niobium superconducting cavity back in 2011, that may have produced up to 10 milli-Newton of thrust with a cavity Q of 11 million. Over the last two years (2015-thru-2016) Guido's team has produced three more niobium cavities, based on the same pill-box beam pipe resonant cavity geometry we tested in the Eagleworks Lab in August 2013 and January 2014, but made from Niobium instead of copper. That has now been tested in the up, down, and sideways configurations relative to the Earth's gravitational field. Guido indicated that these new niobium superconductive cavities that have *NO* dielectric inserts, were now operating with a cavity Q of > 60 million and producing thrust with an efficiency of greater than ∼50 N/kW, with 4 watts of ∼930 MHz RF input. Martin Tajmar mentioned that Shawyer has built a superconducting cavity and this was published in Acta Astronomica [7].

Rodal: It is interesting to note that Fetta has published his superconducting cavity result in the AIAA JPC in 2014 [6] and he was not seeing the efficiency he expected from the use of the superconductor, using helium temperatures around 3 Kelvin. Neither Fetta nor Shawyer are seeing an effect of increase of force proportional to their Q.

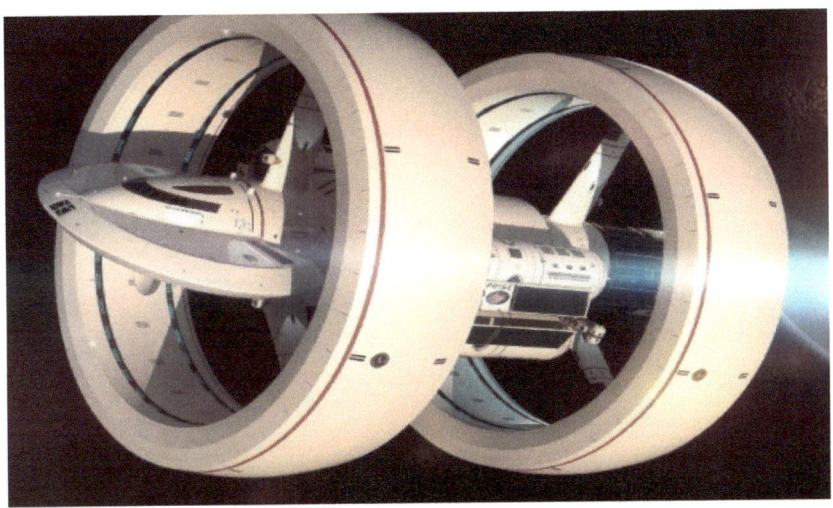

Figure 5: The IXS Enterprise concept design [8], for deep space missions with a FTL, Warp drive engine rings.

Before we move on, let us address the use of warp drives as a possible mode of interstellar space transportation. Sonny White has been pushing this research for some time, in order to promote study in this area. A physicist needs to believe his/her work will be taken seriously by their peers. Many scientists (both physicists and engineers) at this meeting would like to congratulate Dr White for doing an excellent public relations job on their behalf, and for making an attempt to establish gravitational warp theory, as a legitimate field of study. Thank you Dr White!

The Mark Rademaker IXS Clarke concept drawing above is just the interplanetary (solar system) exploration ship since it does not have faster

Figure 6: Pill-box beam pipe design cavity by Guido Fetta. [2]

Figure 7: Three generations of device by Roger Shawyer [9].

than light (FTL) engines on board. The IXS Enterprise [8] on the other hand, has both the EM-drives and warp drive rings. The YouTube video [8] was meant to promote Science, Technology, Engineering and Mathematics (STEM) study in young adults.

2. Two different designs; many experimenters

It is important to note that we have been discussing two very different designs of cavity which have different geometry and different electromagnetic modes of operation. We have mentioned the pill-box beam-pipe design by Guido Fetta [6], and a truncated cone [7,9] by Shawyer, both are illustrated below. There have been attempts to compare the thrust to power input of these devices online. They should not be compared directly, since they are very different in geometry and it would be like comparing apples to oranges.

Both of these models have been tested at the NASA Eagleworks facility. The cavities are seen in the Eagleworks vacuum chamber below. There is a photo of both Guido Fetta and Roger Shawyer with their respective devices.

I wanted to give fair credit to others working in the field of EM-drive propulsion. Here is a list of active scientists that I know of who are currently running their own experimental tests. I give the resonant frequency of their device, name, and location.

- 2.45 & 3.85 GHz, Roger Shawyer ...UK

- 1.937 GHz, Dr. Harold White & Paul March ... JSC/Eagleworks Lab, Texas, USA

- 2.45 GHz, Prof. Dr. Martin Tajmar ... Technische Universität Dresden, Germany

- 24.0 GHz, Paul Kocyla and Jo Hinchcliffe ... Aachen, Germany

- 2.45 GHz, Michelle Broyles ... Colorado, USA

- 2.45 GHz, Phil Wilson ... Australia

- 2.45 GHz, Dave Distler ... Ohio, USA

- 2.45 GHz, Jamie Ciomperlik ... Georgia, USA

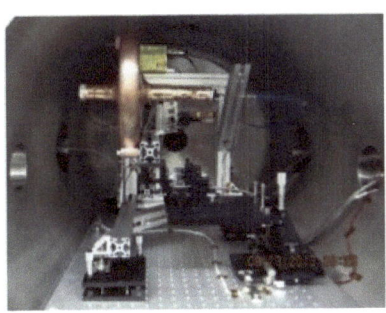 

Figure 8: Pill-box beam pipe cavity in the vacuum chamber at Eagleworks.

Figure 9: Shawyer's truncated cone undergoing testing in the Eagleworks vacuum chamber.

You make the EM-drive out of a highly conductive materials to reduce power losses ($i^2 R$) in the material. The shape is usually a truncated cone of the Shawyer type. These are all AC driven systems, you cannot make a propulsion system with DC power no matter how high the voltage. We have an injection antenna, either a ring shaped one or a small cylinder, which injects RF electromagnetic radiation into the cavity. A variety of modes can be set up inside the cavity, due to the radiation bouncing back and forth between the end plates. You can scale these things to be any size, the resonant frequency would decrease with size, but as Rodal mentioned earlier, the Q increases with size. You can think of these cavities like an acoustic cavity (of a wind instrument), different lengths give you different resonant modes (different frequencies). These cavities are closed, not open ended like an instrument. They must be closed so that the electromagnetic waves can reflect back and forth. The greater the number of bounces, the higher the Q and the greater the electromagnetic field strength within the cavity. The electromagnetic waves will eventually die down due to power (heating) losses, once the RF input is shut off. The higher the cavity Q, the higher the field strength you can achieve inside the cavity and when the RF is shut off, the longer it takes for the electromagnetic field to die down inside the cavity and dissipate as heat.

NASA/JSC Eagleworks Laboratory has a large Stainless Steel & Aluminum vacuum chamber with inner diameter 30" × 38" long . Our RF sources range from 9 kHz up to 2.5 GHz. See AIAA 2014 JPC paper for details [2]. Below in Fig. 10, is a picture of our vacuum chamber, you can also see the rack mount with our data acquisition systems. The vacuum chamber has a sliding tray inside. Fig. 11 shows a photograph of the first generation magnetic damper, using 3 neodymium magnets and an aluminum bar.

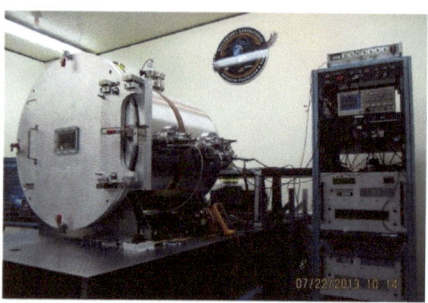

Figure 10: Eagleworks stainless steel and aluminum vacuum chamber, 30" diameter, 38" long. [2]

Figure 11: Magnetic damper on the pendulum balance. [2]

The vacuum chamber has a micro-Newton force resolution-capable torque pendulum (see Fig 12.) The displacement of the pendulum is measured by a Philtec optical sensor, similar to the one Jim is using. The power to the test article is directed through Galinstan contacts (liquid metal) to avoid the weight of the cables producing an unwanted torque on the test device. The procured roughing and turbo-vacuum pumps can pump a clean chamber down to $\sim 5.0 \times 10^{-6}$ torr in about 4 hours, dependent on outgassing. The calibration of the force is done using electrostatic calibration fins, giving a known attractive force (see Fig. 13). A detailed theory of calibration using interlacing fins can be found here [10]. The force from the fins is constant over a small range and has a linear drop-off with distance. The constant force over a small range makes it very convenient for calibration. We used a National Institute for Standards and Technology traceable 1mg mass to calibrate our chemical scale, which then weighed all the test masses for all our calibration runs. We used 200V, 300V, and 400V test pulses for calibration with 3 and 5 kg loads.

You will be reading a great deal about electromagnetic field modes. For example transverse electric (TE) modes inside a truncated cone cavity

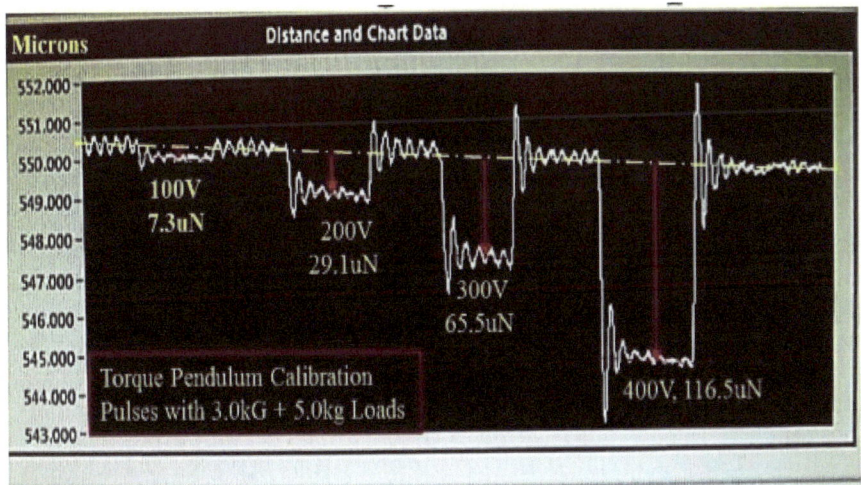

Figure 12: Magnetic damping of the calibration pulses, showing the oscillation of the pendulum. We used 100-400V pulses to calibrate the thrust with loads of 3-5 kg.

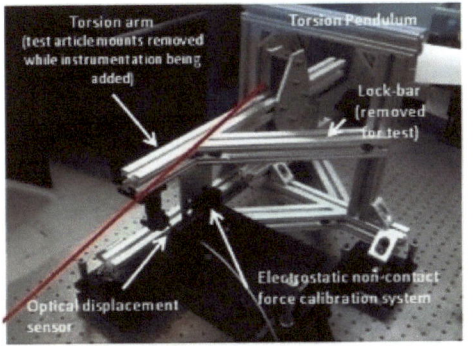

Figure 13: The Eagleworks torsional balance [2].

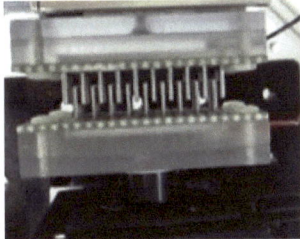

Figure 14: Calibration fins exert a known electrostatic non-contact force on the torsion balance [2].

would look something like Fig. 15. The TE boundary condition is that the electric field at each cavity wall must be perpendicular to the wall or zero there. If the electric field did have a component parallel to the wall, electrons in the conducting wall would flow inside the wall until they produced their own cancelling electric field. The magnetic field cannot have components perpendicular to the wall. The magnetic field must be zero or lie parallel to the wall. So in TE mode the electric field circulates around the z-axis (symmetry axis) of the cone, [11]. In the figures below, the z-axis is vertical, at the center of the truncated cone.

For the TM mode, the situation is reversed. The magnetic field is circulating around the symmetry axis (symmetry z-axis), and the electric field is perpendicular to the wall. See Figs. 16 and 17. The electric field is shown in red, the magnetic field is blue.

The electromagnetic mode plots were calculated by Frank Davies [12] using COMSOL [13] analysis to make the first map of the truncated cavities RF response from 900 MHz up to 2.5 GHz, see Fig. 18. Then we used the Eaglework Lab's Agilent FieldFox N9923A Vector Network Analyzer, to experimentally verify the COMSOL predictions using either its S11 one-port or S21 two-port analysis modes. Frank Davies conducted several finite element analyses of electromagnetically resonant cavities: Cannae's, Shawyer's, Yang's, and Eagleworks' truncated cone, including parametric analysis of different EM-drive geometries. He showed the eigenmodes, eigenfrequencies and S parameters with different graphs, including surface contour plots in 2 and 3-D, and 3-D vector field plots. Davies produced voluminous amounts of data, we don't have room to discuss them here.

Editor's Note: COMSOL is a commercial, finite-element, analysis and simulation software. It has a "multiphysics" capability to simulate different physical phenomena, including electromagnetic radiation.

Bushnell: Where are the Chinese on your list? There was one group headed by Yang Juan, Professor of Propulsion Theory and Engineering of Aeronautics and Astronautics at the Northwestern Polytechnic University in Xi'an. They published a paper in February 2016 [14] where they found their previous positive thrust results to be in error. The university is no longer conducting research into Shawyer's EM-drive. The abstract explains it briefly so I quote: *"In order to explore the thrust performance of microwave thruster, the thrust produced by microwave thruster system was measured with three-wire torsion pendulum thrust measurement system and the measurement uncertainty was also studied thereby judging the credibility of the experimental measurements. The results show that three-wire torsion pendulum thrust measurement system can measure thrust not less than 3 mN under the existing experimental conditions with the relative uncertainty of 14%. Within the measuring range of three-wire torsion pendulum thrust measurement system the independent microwave thruster propulsion device did not detect significant thrust"*. It appears the power leads to the cavity were heating up and thermal expansion of the leads was responsible for giving a force which was mistaken for thrust from the cavity. She had

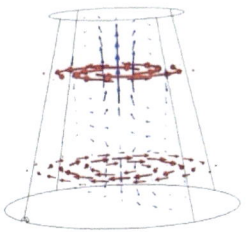

Figure 15: The transverse electric mode, (TE012) electric field is red, magnetic field is blue. Frequency 2.1794 GHz, [12].

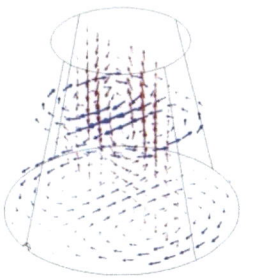

Figure 16: Transverse magnetic mode (TM112), electric field in red, magnetic field in blue. Frequency 1.9355 GHz, [12].

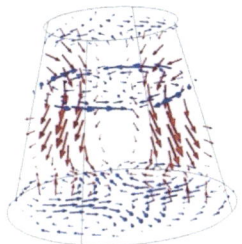

Figure 17: Transverse magnetic mode (TM212), electric field in red, magnetic field in blue. Frequency 2.4575 GHz, [12].

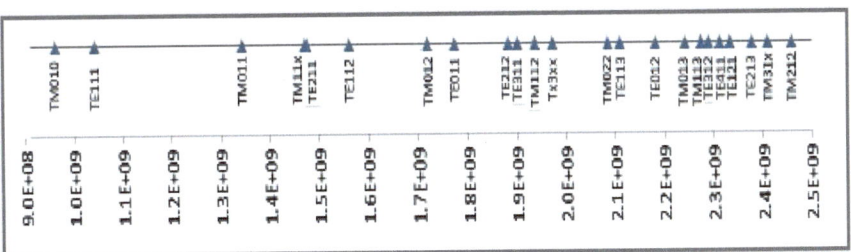

Figure 18: Spectrum of electromagnetic field modes in the truncated cone cavity from 900 MHz to 2.5GHz, [12].

about 30 Amperes of current flowing through these power leads, which causes significant heating. Prof. Yang's stance on the EM-drive is that the thrust she published in previous papers were experimental artifacts due to the power cables. When she used a battery she measured no thrust. Her official bio-sketch at her new University no longer features her research on Shawyer's EM-drive among her "selected publications". Instead, her research is now on the conventional, classic Microwave Plasma Thruster, which uses a propellant for thrust.

Continuing now, there are a few others who have published their findings that I am aware of:

- Kurt Zeller and Brian Kraft, California Polytechnic State University, San Luis Obispo, [15].
  https://www.linkedin.com/in/kurtwadezeller

- Iulian Berca, Romania. The first citizen scientist to report an independent test
  http://www.masinaelectrica.com/emdrive-independent-test/

- Eugene Samsonov, obtained no thrust, but the only citizen scientist to use a battery to conduct his tests, and hence his test was not subject to power cords issues, (thermal or electrical). http://vixra.org/abs/1603.0153 & http://vixra.org/pdf/1603.0153v1.pdf

- Sorry if you have been missed out... there are many of you out there!

Samsonov's truncated cone had a loaded Q-factor of approximately 3100. With his TE012 resonance mode, with an input power $\sim$30 W, it was very unlikely he would see anything to begin with due to the thrust scaling as the electric field strength to the 3rd or 4th power. It took 30 W with a loaded Q-factor of $\sim$23000 in the Eagleworks' copper frustum's TE012 mode, plus the High Density PolyEthylene (HDPE) discs, to get a thrust signature of $\sim$65 μN towards the small diameter end of the cavity. Let's compare these two experiments. In terms of power P in watts, Samsonov had a $P \times Q$ product of $30 \times 3100 = 93000$, whereas the Eagleworks lab's TE012 experiment had a $P \times Q = 30 \times 23000 = 6.9 \times 10^5$.

So Samsonov's thrust levels should have been approximately factors of
$(93/690)^3 = 2.4 \times 10^{-3}$, or $(93/690)^4 = 3.3 \times 10^{-4}$ less, than the Eagleworks 65 μN TE012 test. Or, it should have been between 0.16 μN <

thrust < 0.02 µN, assuming the Eagleworks' HDPE discs played no factor in the size of the thrust. (Jim's Mach-Effect conjecture would argue otherwise).

Now take into consideration that the battery powered Eagleworks Cavendish Balance free-flyer tests using the same copper frustum with HDPE discs demonstrated that the DC power source for the RF amplifier makes no difference in the thrust production in these devices. This indicates to me at least that the Samsonov test was a nice try, but it did not get the test article up into the proper $P \times Q$ power product range to see the effect using Samsonov's force detector with an as-built noise platform of 13 µN per micron of displacement while in a 1 atmosphere air environment.

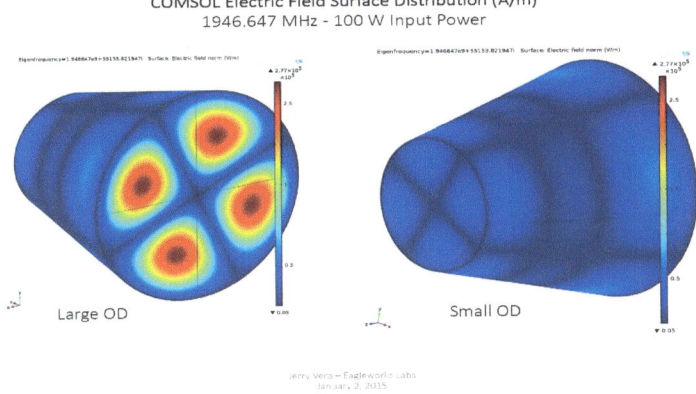

Figure 19: COMSOL simulation of the TM212 mode, showing the electric field strength on the surface of the truncated cone cavity. Frequency 1.946 GHz, at 100 W power[12].

Regarding the electromagnetic modes, Fig. 19 shows a contour plot of the electric field strength at the surface of the truncated cone for mode TM212. Fig. 20 shows a similar contour plot over the surface of the cavity but this represents the magnetic field strength. Fig. 21 shows a calculated energy-dissipation (temperature) plot, made using COMSOL. Over the large diameter end of the cavity and next to it, is an infra red camera photograph of the large diameter end of the cavity. There is a striking similarity between these two images in Fig. 21. It is important to know exactly what mode you have excited, inside the cavity, in order to predict the thrust values you can achieve. This is *the first time* an experiment using an IR camera has verified the exact mode of electromagnetic excitation within the cavity. Eagleworks was the first group to verify experimentally exactly what mode they had excited in the truncated cone.

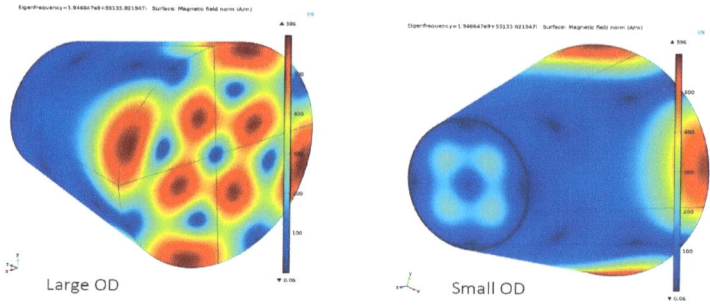

Figure 20: COMSOL simulation of the TM212 mode, showing magnetic field strength on the surface of the truncated cone cavity. Frequency 1.946 GHz, at 100W power. Compare with the thermal image in Fig. 21.

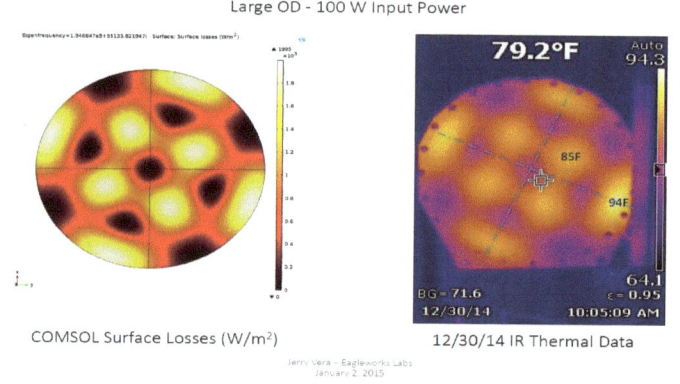

Figure 21: The figure on the left is a COMSOL energy loss plot (heating) showing the large diameter disc of the cavity. The right figure shows an infra red image of the large diameter disc of the cavity. Note how the hot spots match the high energy loss regions in the simulated image on the left.

The eigenfrequencies that were derived in the COMSOL study, by Frank Davies, were verified by the impedance resonances for the cavity detected at various frequencies (between 900 MHz and 2.6 GHz) using a Fieldfox, Vector Network Analyser. The agreement was very good between the theory and the experimental observations. The Agilent Technologies VNA model N9923A is seen in a photograph Fig. 22, the screen shows multiple impedance resonances. In Fig. 23. we show the com-

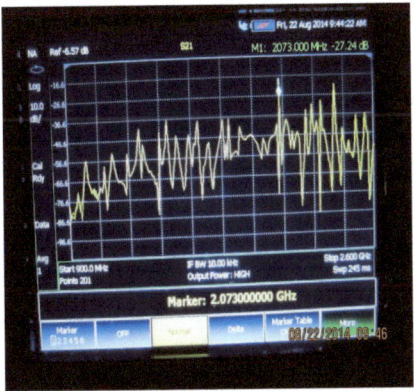

Figure 22: Agilent Technologies Fieldfox Vector Network Analyser, model N9923A, showing impedance resonances in the truncated cone cavity at Eagleworks. Note the frequencies are between 900 MHz and 2.6 GHz, which corresponds to the frequency range studied by Frank Davies using COMSOL.

parison between the VNA measured spectrum and the calculated results from the COMSOL study.

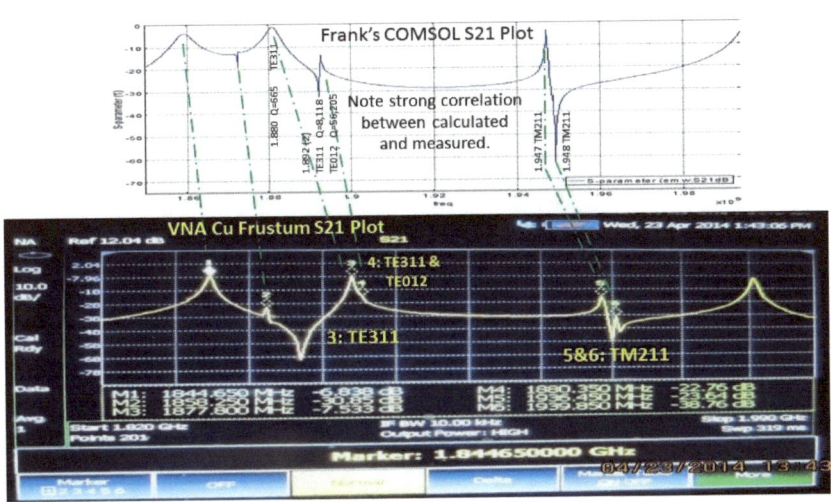

Figure 23: This is a comparison between the eigenfrequencies calculated by Frank Davies using the COMSOL simulation software and the measured response of the truncated cavity using the Fieldfox VNA.

Fearn: Is the shape critical? Why don't you use something simple, like a cylinder or an ellipsoid, something you can solve for the cavity modes analytically? Why this weird, chopped-off cone shape?

March: First off, the cavity has to be an asymmetrical shape to give a force in one direction or another. A symmetric shape would not work. Otherwise you would have equal forces on the end caps and they would cancel and so no net force.

Rodal: Some of the theories have a polymer insert at one end, so if you put a HDPE disc at one end of the cylinder, that would act as the asymmetry and then the forces would not cancel, and so the cylinder should work. Other theories have a paramagnetic material at one end which acts as the asymmetry, or a different form of dielectric, all these are different forms of asymmetries.

March: Yes, when you put in an insert (polymer disc, dielectric, or paramagnetic material) that would cause an asymmetry and that may well open up the use of cylinders and ellipsoids and other shapes. I personally have not done any experiments using a cylinder with dielectric disc at the end, so I have no direct experience with it.

Broyles: There was a group of college students in California [15], who tried to use a cylinder with a dielectric insert, but they were having some problems with their test setup so, at the moment, their results are inconclusive. What we are doing now is going with a slightly different cylindrical shape and repeating the experiment and trying variations of the truncated cone shape, with a polymer disc insert towards the smaller end.

Cole: Is the acceleration always toward the small end of the truncated cone?

March: It can be in either direction depending on the mode that is excited within the cavity.

Christie: Is the length of the cavity half the wavelength of the resonance frequency?

March: The length of the cavity is either half the resonant wavelength or a quarter wavelength, depending on how you set it up.

Christie: So if you are heating the side walls, they are not exactly at 1/2 or 1/4 wavelength spacing, is that going to effect the Q or the dissipation?

March: Yes the wall angle is part of the optimization process. I would like to have made a whole family of cones with different angles and done a systematic test of all of them to see which gave the most thrust, but we never had the time or rather budget for that.

Meholic: Did Shawyer ever give any indication that he did any of those types of studies?

Rodal: Shawyer had a patent from 1988 for a cylindrical cavity with a dielectric cone inside. He went from that to a truncated cone cavity with and without a polymer disc at the small end. There is no patent for a cylindrical cavity with asymmetric ends. Shawyers explanation for the acceleration makes no sense. He is only using Maxwell's equations and special relativity, we know from conservation of momentum, in that case, that there should be no thrust. Shawyer claims that there is no pressure on the side walls, he claims that the only pressure is on the flat end plates, which is not physically possible.

Meholic: He gives no proof, just states there is no pressure on the side walls?

March: Yes. I think we all agree that Shawyer's theory makes no sense, but we also agree there is an acceleration, so we need to find a better theory.

3. Propellant-less propulsion testing, before and at Eagleworks

I got into this business because of Dr. Woodward. I ran across a paper of his back in 1988 and started pursuing propellant-less propulsion from that point. My first build was in 2004, it was a Mach Lorentz thruster. (See Fig 24 & 25). The thruster consisted of a small ring of capacitors that had a toroidal magnetic winding around it. It was driven with an open wire transmission line system that drove something like 800 V peak across the capacitor ring and across the inductor (they were wired in series). There was a $\lambda/4$ phase shift between the capacitors and the inductor. With a 2 MHz frequency voltage, I saw some interesting results. I measured about 2 to 4 mN of force generated. This result was reported at "The Space Technology & Applications International Forum" (STAIF) in 2006. I never could replicate that, when I went to a co-axial version of it, but I never got to the same peak voltages either. When I went to a totally enclosed system with Faraday shields and all the rest, I could only drive up to, with the same RF frequency, about 160 Volts, and I didn't seen any significant force at those voltage levels.

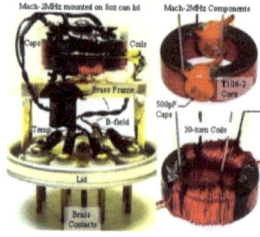

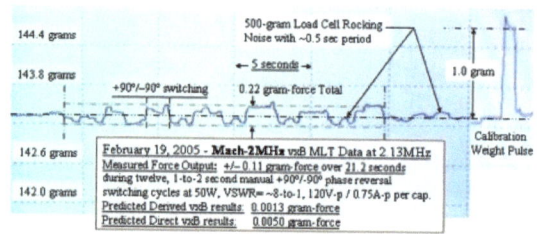

Figure 24: Shown is the 2004 test Mach Lorentz Thruster (MLT). It consists of a small ring of capacitors with a toroidal magnetic winding.

Figure 25: The 2004 test results for the MLT, using 2 MHz frequency and 800 volts. The force was about 2160 μN.

The second test article was actually Jim's device, that he asked me to test for him. It was ~ 1 mN at most. This is also reported in the STAIF 2006 report.

I had been laid off from the Orion program at JSC and Sonny White brought me out of premature retirement to help develop his new EagleWorks lab. This was back in May 2011. The objective was to test primarily EM-drives. So we got the vacuum chamber sitting on a floating optical isolation table to minimize vibration. We got the vacuum pumps and the usual lab equipment which took about 18 months total. Our torsion pendulum uses two flexural bearing blocks, which can support up to 100 pounds [16]. This is the same bearing as used by Nembo and Martin Tajmar.

Sonny White tested his resonant cavity on November 20 2012. (See Fig 26 & 27.) At a resonant frequency of 4.14 MHz , we applied a power of 150 W of radio-frequency. The straight thrust prediction, for this applied capacitor-ring voltage level, was about 1.5 μN. We observed a force of 2.6 μN. I'm fairly sure I messed that up.

Eaglework's first outside test was for a DARPA customer, during the January 2013 to June 2013 time period, testing an electrostatic device (Scorpion Thruster) from Gravitec (founded by Hector Serrano). See Fig 28 & 29. We primarily observed electrostatic interactions with the vacuum chamber, but did find transient thrust pulses $\sim$ 110 μN with a switching, high voltage power supply (30 kV). This was after an extensive shielding campaign was accomplished.

Eagleworks second customer test series was for Cannae LLC (Guido

Figure 26: Sonny White's resonant cavity "Q-thruster"

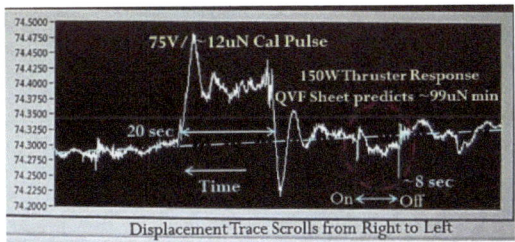

Figure 27: The Q-thruster at frequency 4.14 MHz and power input 150 W. Driving at 22.9 volts per capacitor gives a force of 2.6 μN.

Figure 28: DARPA test article 2013. From Gravitec, work of Hector Serrano.

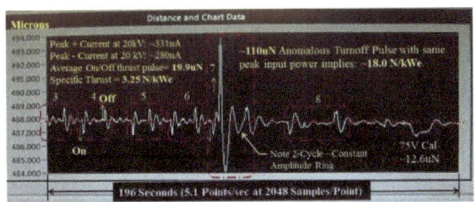

Figure 29: Test results for the DARPA article, showing an anomalous transient thrust pulse with high voltage power switching.

Fetta) and his 937 MHz, TM010 pillbox resonant cavity thruster in August 2013 and January 2014.

Mr. Fetta brought two kinds of pillbox, or "pancake", test articles, one with machined radial slots on the interior of one wall in the cavity, and a "null" test article that had no machined slots. However both test articles used a 1.0 inch outer diameter by 1.63 inch long Teflon (PTFE) cylinder in their RF power input section. The input RF power section was just over 5 inches long. There was a 0.115 semi-rigid coaxial cable with SMA attachment used for the input power. The main cavity was just less than 11 inches in diameter and about 1.5 inches wide. The RF signal stub antenna was in the opposite pipe from the input RF (about 3 inches long) and that section was used as the output to the spectrum analyzer.

A reversible thrust signature in the 35-to-65 μN range was observed during the August 2013, ten day test period when its Z-matching Teflon cylinder was mounted in the throat of the pillbox cavity's RF feed line. The loaded Q factor for the slotted cavity was 8500, the cavity without the slots had a loaded Q of 9500, which makes sense because the slots just give more surface area for $i^2 R$ losses.

During the Cannae test setup, it was determined that the Eagleworks' liquid metal contact array used for passing RF or DC power, control, and signal lines from the test article to the outside world severely attenuated RF signals above $\sim$ 4 MHz, which made them unusable for passing UHF and above RF power signals through them. Above 50 MHz we couldn't get anything more than a couple of watts through the liquid metal contacts, so for UHF the Galinstan contacts were only used for DC voltage.

The solution to this problem was to mount Fetta's 937 MHz, 30W RF amplifier in the vacuum chamber and use it, and its heatsink, as a counter mass for the torque pendulum's test article. The Eagleworks lab then used this approach for all future testing. Later, we built an integrated copper frustum test article that we had to marry to its RF amplifier, needed to get a symmetrical forward and reverse thrust response from the torque pendulum. The amplifier had electrolytic capacitors inside it that were not vacuum rated, so we could not run this device in a vacuum. (You need ceramic capacitors for a vacuum so we ran those tests later with a different amplifier).

Tajmar: What happened to your liquid metal contacts as you went to UHF ?

March: The liquid metal heated up a little with DC, we measured it with an IR sensor. For RF they got warm. There was too much cross talk between all 10 liquid metal contacts, I should have had a co-axial arrangement of contacts, there was no coaxial shielding.... lessons learned. That's why we ended up making them all DC.

We can measure what the electric field strengths are inside the cavity using a small ring antenna probe. Thus we can confirm the fundamental transverse magnetic mode, TM010 mode structure. The magnetic field circulates around the perimeter of the saucer shape whereas the electric field is axial. The RF is injected in from the right side beam pipe and is taken out to the spectrum analyzer from the left side pipe, see Fig. 8.

Mr. Fetta brought two more TM010 test articles to the Eagleworks lab for testing in January 2014. One of these "pancake" test articles had internal radial slots and the other one did not, however unlike the previous two "pancake" resonant cavities, neither of these two new test articles had Teflon cylinders in their RF input impedance matching section. No detectable thrust signatures were observed with these two new pancake test articles without the PolyTetraFluoroEthylene (PTFE or "Telflon") inserts for "impedance matching".

The third Eagleworks customer test series was for Roger Shawyer's EM-drive. Shawyer likes to use the TE modes. This replication effort used a truncated copper cone, fabricated in-house, loosely based on Shawyer's 2nd generation dynamic thruster design. Sample data is shown below in Fig. 30.

During this time period, approximately a six month long program of COMSOL simulations, examined this frustum's various resonant modes. This work was done by the Eagleworks COMSOL analysts, Frank Davies and Jerry Vera. See the detailed plot of the resonant frequencies and the electromagnetic modes that can be excited at those frequencies in Fig. 31.

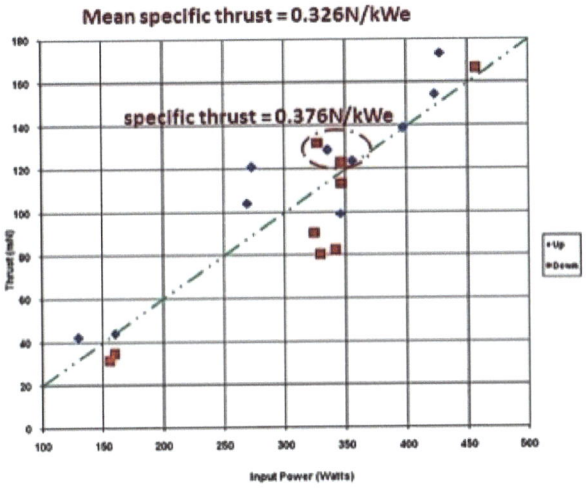

Figure 30: The data for Shawyer's 2nd generation device. The triangles are for upward motion, the squares are for downward motion. The vertical axis is thrust in mN, the horizontal axis is input power in watts.

Magnetrons are notoriously wide bandwidth, about 3 MHz or so. Shawyer went from wide bandwidth to narrow using a travelling wave tube, which has a bandwidth of 10 kHz or so. Shawyer built three types of test article (see Fig. 7). The first cavity had flat end caps and produced about 16 mN of thrust. The 2nd generation and everything after used spherical end caps, like mirrors in reflecting telescope, Fig. 32. When you use the spherical end caps the electric fields increased by a factor of 5-10. If the thrust goes like the electric field cubed, that is a huge improvement. We are seeing microNewtons and he is getting milliNewtons. We should point out that Shawyer has never reported in-vacuum testing, all his tests are in air so there is a thermal component to it. Shawyer knows about buoyancy and has corrected for it. You can rotate your cavity 180 degrees and subtract one thrust from the other, since the buoyancy will remain the same, it subtracts out. Also the buoyancy effect is small, ~3 mN with respect to the expected thrust signature ~20mN.

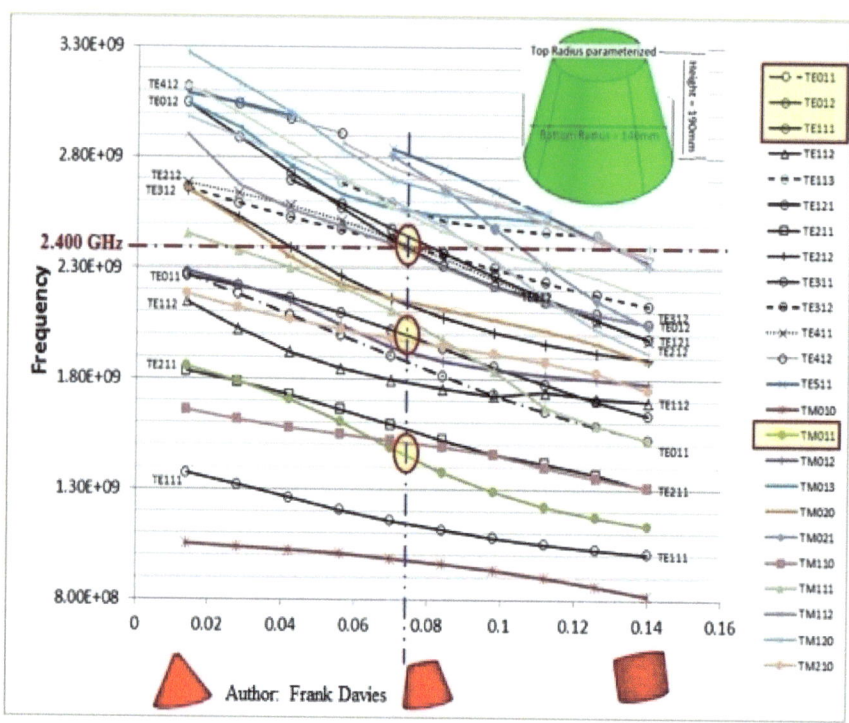

Figure 31: COMSOL data provided by Frank Davies. Modes for a given frequency are plotted. Using an empty cavity, the geometry was changed from a cone on the left to a cylinder on the right. All the modes were noted on the right side. Eagleworks decided to test a geometry in the middle. The effect of the HDPE disc was to downshift the resonant frequency of the cavity by a certain percentage.

Fearn: This may be a dumb question but does it make any difference where the RF input feed is, in the middle of the cavity or at the end?

March: Oh yes, it makes a difference. The optimum place to put the feed is in the middle. So it echoes. You want equal distance from the antenna to the end plates, otherwise there can be destructive interference between the wavefronts.

The third generation device was paid for by Boeing. It weighed 2.92 kg, was 265 mm diameter at the base plate and a height of 164 mm. This also had spherical end caps. The mean specific trust claimed by Shawyer was 0.326 N/kW. He used a 1/2 loop antenna tuner in this 3.85 GHz room temperature EM-drive. Further details can be found here http://emdrive.com/flightprogramme.html. The Boeing guys said they got nothing, they could not get the cavity to resonate.

The Eagleworks copper truncated cone (frustrum) build-up project was started in the October 2013 time frame, with the fabrication of our current copper frustum that was loosely based on Shawyer's 2nd generation dynamic test copper cavity. (Eagleworks prefers to use the TM modes). The COMSOL code was used (summer 2013) to verify that according to Maxwell's equations alone there is no thrust on the cavity due to internal radiation pressure, as expected. So in order to account for the force on the cavity, new physics would need to be applied, not just Maxwell's equations.

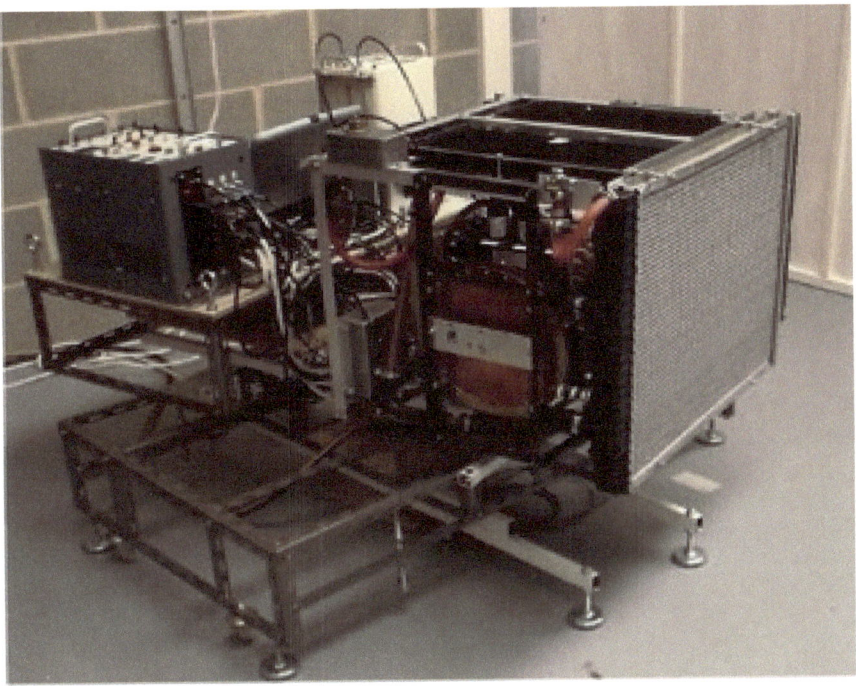

Figure 32: Shawyer's 2nd generation device on a rotating test bed. This cavity used spherical end caps. For the Dynamic test, a thrust of 96 mN was recorded for an RF input power of 334 W or 0.287 N/kW.

It was decided to build a fixed-geometry frustum configuration and then electronically tune the solid state, narrow band RF signal frequency source to the *now* resonant frequency in question using a hand tuned and then a phase locked loop frequency tracking method. Construction was completed in December 2013.

During the spring and summer of 2014 we explored a number of this copper frustum's RF resonant modes from its fundamental TM010 mode through its other TE and TM modes up to 2.50 GHz. The affect of the HDPE discs was that the Q went down from 40,000 with no dielectric disc to 25,000 with the disc present. Also, the resonance frequency of the cavity would decrease by a certain percentage with the HDPE disc present; in the case of one cavity we built, from 2167 MHz down to 1880 MHz.

During the spring and summer of 2014, March explored a number of this copper frustum's RF resonant modes with and without dielectric disks from its fundamental TM010 mode at 957 MHz up through its other TE and TM modes up to 2.5 GHz.

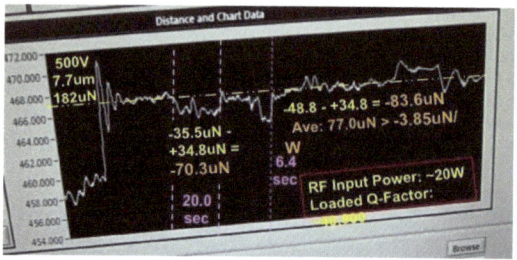

Figure 33: First run of the Eagleworks EM-drive, based on Shawyer's 2nd generation model. No disc present. Thrust to the right as shown by the red arrow.

Figure 34: First data plot for our EM-drive. Note the rather large test pulse of 500 V. The loaded Q was 40900.

In this frequency range we found that the TE012 mode at 2167.14 MHz had the narrowest -3dB bandwidth and thus became our first resonant mode to be used for thrust production. See Figs. 33 & 34. During our initial test at 2167.14 MHz, we found that with $\sim$ 20 W of RF input power, the average thrust signature was -77.0 μN after correcting for the actual calibration magnitude and the 28V dc offset bias of 34.8 μN. Thus the TE012 thrust efficiency = -3.85 μN/W. The TE012 copper frustum test without dielectric yielded an average - 77 μN negative going signal with $\sim$ 20 W of RF (- 3.85 μN/W), in the direction of the large end of the Frustum.

Williams: I thought you had said in your 2014 paper, that when you had no HDPE discs present there was no thrust?

March: Yes, I had misinterpreted the size of a calibration pulse, I thought it was 29 μN and it was actually 187 μN. That changed my analysis of the data and the value of the recorded thrust. So yes we did see a -77 μN thrust with no discs present.

Rodal: I thought that Shawyer had claimed that the thrust he saw, with the same configuration (no HDPE discs), was in the direction of the small diameter end?

March: Yes he did claim that, and I cannot reconcile that discrepancy.

Thinking of Dr. Woodward's Mach-Effect electrostrictive work, I also tried the same copper frustum cavity with two, 6.13 inch by 1.06 inch thick polyethylene (HDPE) disks mounted at the small end of the frustum. I noted that its TE012 resonant frequency was now down at 1880.62 MHz with a loaded -3dB Q-factor of about 25000 and -3d B bandwidth of 88.0 kHz, which made for easier manual tuning. See Figs. 35 & 36. Measured thrust level at this new TE012 resonant frequency with the same 18.7 W of RF was now a positive going 37.3 μN, or a thruster efficiency of 2 μN/W.

Figure 35: Eagleworks EM-drive, using 2 HDPE discs at the narrow end. Thrust to the left as shown by the red arrow.

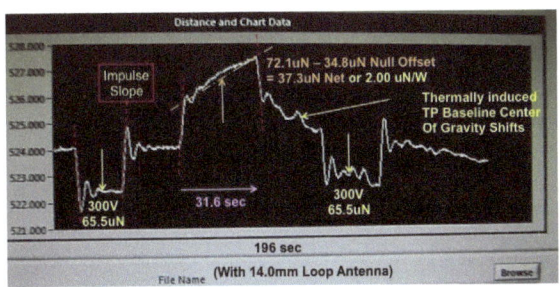

Figure 36: Plot for our EM-drive using 2 HDPE discs at the narrow end.

Copper oxidizes in air, especially in Houston with all the humidity, so after the copper is rolled out we polished it. Here are the details: The process for finishing the interior of the copper frustum was that David Fletcher polished the interior surfaces of the copper cone and 1.0 oz copper (~35 micron thick), single layer PCB end-plates with buffing compound with a power drill buffing wheel, then a warm soapy water wash, followed by distilled water rinse and dry cycle. He did not indicate to me that he used any finishing polish afterward. I then spray coated all the copper frustum's polished surfaces with one coat of MG-422B silicone conformal coating [17]. This silicone conformal layer was probably around 1.0 mil (0.0254mm or 25.4 microns) thick that was then cured in an oven set at 150 Fahrenheit for 90 minutes. That works very well to prevent oxidation.

I used PC boards for the end plates of the cavity, these had 1 ounce of copper on the board. This allowed me to use thermal imaging at the ends and verify the electromagnetic mode structure inside the cavity, which would show up as heat. The cavity had a 14.8 degree cone slope. The total weight of the cavity without the HDPE discs was 1.6 kg, and with the 2 discs was 2.57 kg. I had 6.125 inch diameter by 1.0626 inch thick discs of HDPE. I tried 1, 2, and 3 discs, in the cavity. The optimum seemed to be 2 discs for thrust production. The discs were held in place by teflon or polypropylene bolts. Nylon bolts tend to melt in RF fields. We did try Teflon tape around the nylon screws, it didn't work so well. I drilled and tapped the discs and used cap screws.

Our 2014 and early 2015 work was all in air, since we did not have an amplifier that could go inside the vacuum chamber. We got a vacuum ready amplifier in the summer of 2015.

We used magnetic loop antennas to excite the modes in the cavity. One half a millimeter made all the difference with these loops. See Fig. 37. I have not yet tried a loop at the dead center of the cavity. Ours was mounted about 15% up from the bottom. This small loop would not have been deep enough into the cavity to be along the symmetry axis. Deciding on where is the best placement of the antenna, depends on what mode you are trying to excite. Having a loop antenna horizontal in the center of the cavity would work for the TE01X modes, that have E-field donuts wrapped around with magnetic toroids. Then you would raise and lower the loop for impedance matching.

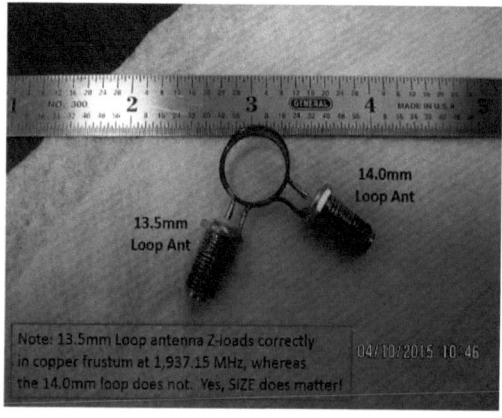

Figure 37: Magnetic loop antenna, the difference of 1/2 mm made all the difference.

I was manually tuning the system throughout most of this work. At the TE012 mode, the loaded Q-factor was approximately 25000. The bandwidth was less than 100KHz. I was trying to hand tune the resonant frequency 1.88 GHz with a low 100KHz bandwidth and I just couldn't dial it in and track it. As the cavity heated up and expanded I couldn't keep the cavity in tune. The TE012, TE013 modes that Shawyer used have a higher Q but that also means that they have a narrow bandwidth and they are hard to excite initially, and harder to maintain as the cavity heats up, especially if you are manually tuning the cavity.

The TM212, at 1937.6 MHz, was easier for me to dial in by hand. It also seemed to produce a little more thrust with the HDPE discs in there. The quality factor went from 25000 to a -3dB loaded Q-factor of around 7000 and that means the bandwidth opened up to about 314 KHz. It was much easier for me to keep track of the resonant frequency and keep the cavity in tune with the RF input. We tried a phase locked loop for the frequency stabilization, designed by the NASA/JSC/EV electronics group on a volunteer basis, but that was not optimum since the mode structure could change. Typical thruster output was 90.4 μN with an thruster efficiency of 4.54 μ N/W, utilizing the Class-A, ZHL-32W-252 Mini-Circuit RF amplifier.

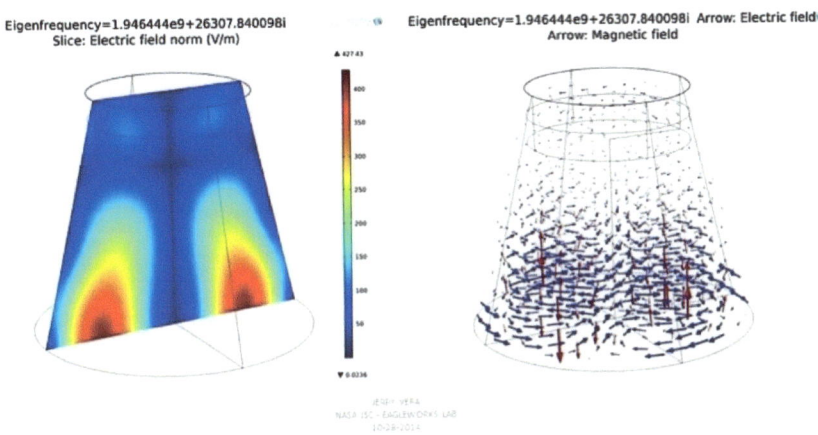

Figure 38: Eagleworks copper cavity, TM212 mode E&M field distribution and strength. On the left you see a contour plot X-section of electric field strength in the center of the cavity. On the right is a vector field distribution, electric field is in red arrows, magnetic field is in blue. *By Jerry Vera, 28th Oct. 2014.*

The best way to go, by far, with automatic tuning is with a VSWR minimum tracker, which tracks how much power is reflected back from the cavity at the input. You want the power reflected back to be a minimum always.

Williams: What do these mode numbers stand for?

Rodal: There is no convention for mode shape numbering for a truncated cone cavity. The numbers are based on a cylindrical cavity. The TE and TM refer to transverse electric and transverse magnetic modes, which we have explained earlier. The first number has to do with the azimuthal direction, the second number represents a radial direction (normally perpendicular to the surface of the cylinder), and the last number is along the z-axis of symmetry, along the center of the cylinder.

Tuning the cavity is a multi-staged affair. The following descriptions are somewhat verbose, but I hope you find them useful:

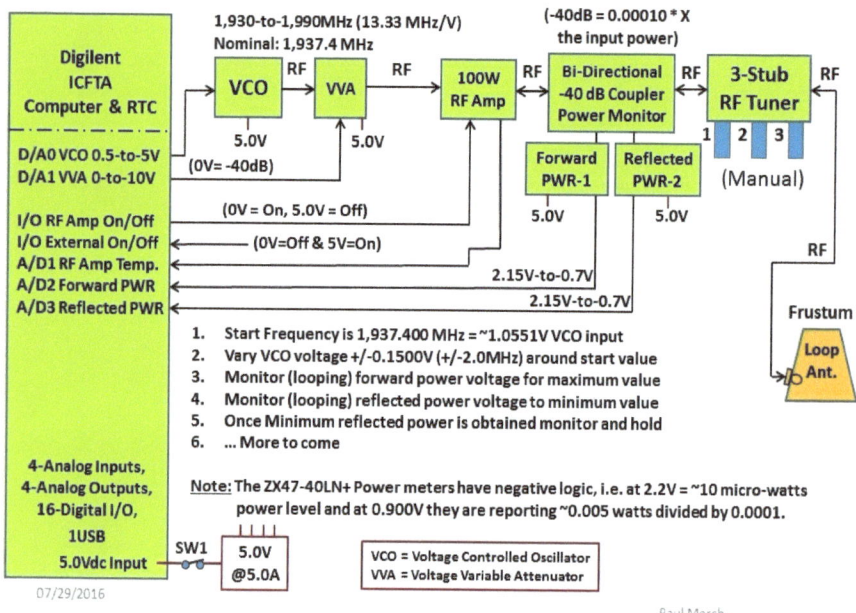

Figure 39: ICFTA tuning subsystem block diagram.

- At the EW-Lab we had Frank Davies or Jerry Vera run a eigen value COMSOL analysis of the frustum in question that provided the frequencies for the resonant modes over the selected frequency range from the cavity's fundamental or lowest frequency TM010 resonant frequency. See the attached plot.

- After picking the resonance mode we want to excite, say the TE012 mode, I would then use the lab's Agilent FieldFox Vector Network analyzer in its S21 two-port configuration using the frustum's main RF input port antenna. I use its field sense port antenna to acquire the actual resonant frequencies for the modes of interest. I would then compare the VNA plot over the specified frequency range to the COMSOL analysis and validate what the actual resonant frequency is per the VNA S21 plot.

- Next I would insert the FieldFox VNA at the output of the RF amplifier going to the frustum to tune the RF amplifier's 50 Ω impedance matching network utilizing the narrowest frequency sweep bandwidth for the resonance in question. The 50 Ω impedance matching network consisted of the copper frustum's main RF input rotatable loop antenna, the transmission line 3-stub tuner and all the coaxial cable and connectors in between the frustum and the RF amplifier. That would entail recording the VNA S11 response plot, Smith Chart, and phase responses by first using the rotation of the frustum loop antenna. I would then repeat this process utilizing the 3-Stub tuner and then iterate between the loop antenna and 3-Stub tuner until I obtained the lowest overall S11 minima for the frustum resonance of interest. See attached two summary VNA slides with the frustum loop antenna and RF amp's 3-Stub tuning solutions for the TM212 mode.

- Lastly I would then vary the Phase Locked Loop Voltage Controlled Oscillator input control voltage to vary the RF amp's frequency, (control pot on the Frustum control panel), to first match the previously recorded VNA resonant frequency. I would then monitor the forward and reflected RF power meters attached to the RF amp's 50 Ω -30 or -40 dB dual directional coupler's forward and reflected output ports via our LabView control panel, to continuously minimize the reflected RF power coming back from the frustum, which at the same time maximizes the forward RF power.

- In our latest version of this tuning business, needed for the Cavendish Balance test, my semi-continuous tuning inputs were replaced with a micro-controller programmed by Sonny, that would automatically keep the RF reflected power at a minimum value by dithering the VCO output frequency around the frustum resonant frequency.

- The next step in automating this Frustum tuning procedure is to add a rotary actuator to the frustum loop antenna and three linear actuators to the 3-stub tuner, so all these tuning steps can be controlled by the micro-controller while is strives to maximize the forward RF power while minimizing the RF reflected power. However, we have noticed that due to the resonant phase shift requirements of the Mach-Effect wave equation, we will have to provide a frequency offset to the above max/min RF power solution that will maximize the thrust production utilizing an onboard the frustum or MEGA drive accelerometer signal as this algorithm's main input.

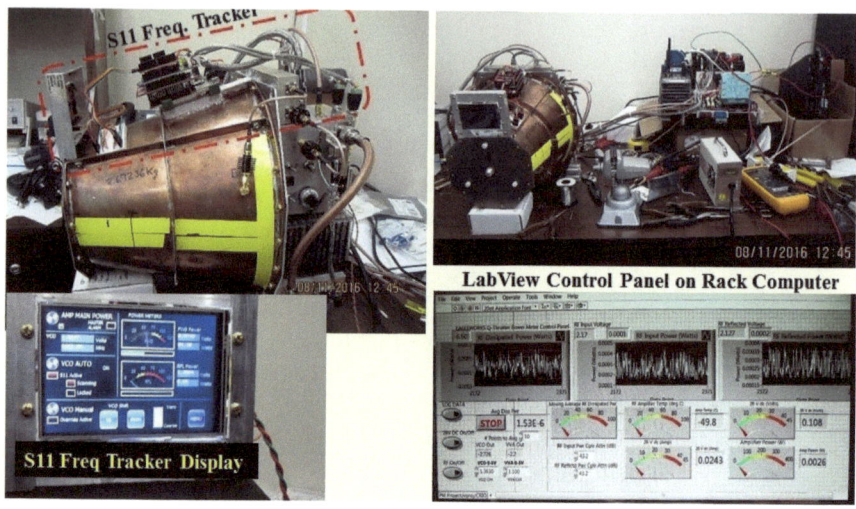

Figure 40: Eagleworks Lab. ICFTA with S11 frequency tracker photo.

The first outside Eagleworks Lab Report paper was published and presented at the the 50th AIAA Joint Propulsion Conference (JPC) in late July 2014 [2]. In July 2014, NASA called a Blue-Ribbon panel of eight PhDs that were asked to evaluate Dr. White's Quantum Vacuum Conjecture (QVC) and its associated experimental test program. Their conclusions were that they thought that the QVC was either "profound" or just a "mathematical coincidence". They were less critical of our experimental program. In August of 2014 DARPA had the JASON Group interview Dr. White in CA on his QVC conjecture. They did not like it. The first group *Dynamics of the Vacuum* theory paper was published in November of 2014, demonstrating that the QVC was NOT a mathematical coincidence, with a follow-on *Characteristics of the Vacuum* paper in 2015 by Dr. White [3]. Whether the QVC is "profound" or not, is yet to be determined.

During the winter and spring of 2014-2015, Eagleworks lab performed a set of in-vacuum test runs, with a split copper frustum, with PE disks and RF amplifier system that generated very clean thrust pulses in one direction, but almost non-existence thrust pulses in the other direction when the copper frustum was physically reversed in its torque pendulum mount. The problem appeared to be centered on having to change the system's physical mass configuration when doing so. Solution was to integrate the copper frustum with its RF amp and RF plumbing into one package, so RF cabling would not change when the test article was reversed. See Fig. 41.

The drawback to this integrated test article approach was that it doubled the sprung mass "flying" on the torque pendulum because in the original split configuration, the RF amplifier was being used as the counterbalance mass for the frustum test article. The results of doubling the sprung mass was a major increase in the torque pendulum's force resolution noise platform and a slowing of its dynamic response. However during the fall of 2015 Eagleworks performed an integrated copper frustum test article (ICFTA) in-vacuum ($\sim 8 \times 10^{-6}$ torr) test series in the forward, reverse, and null thrust test orientations. See Fig. 42. Thrust levels in the 40-to-120 μN levels with up to 80 W of RF input were observed, but the thrust traces were contaminated with thermally-induced center-of-gravity shift artifacts in the torque pendulum [18].

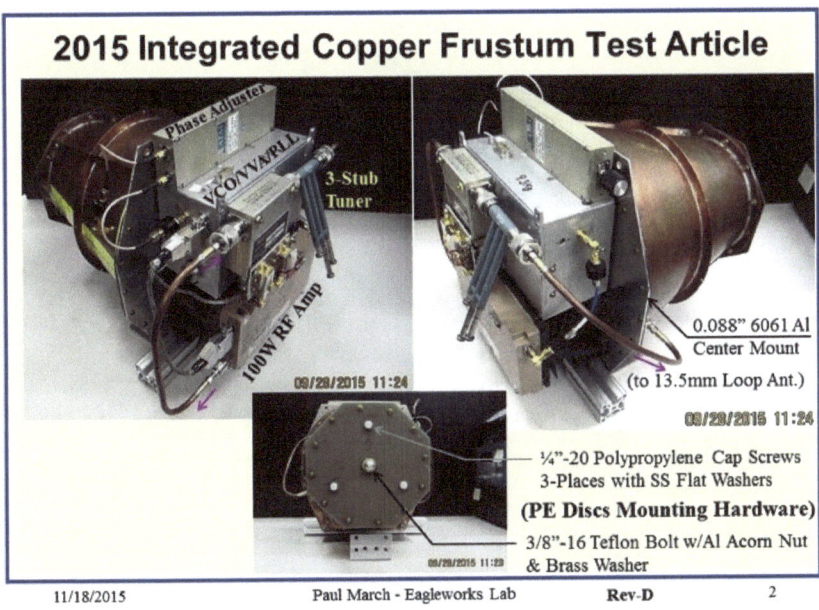

Figure 41: Eagleworks copper cavity, with the power system onboard.

Tajmar: Can you elaborate a little on the Lorentz force you get from the wires going to the test article?

March: Yes Martin. We've got a twisted pair wire coming from the liquid metal contacts going over to the RF amplifier. For this amplifier it was about 5.5 Amps and 28 Volts DC. If you had a single wire, going through ground return you would have a huge area in the loop which could interact with

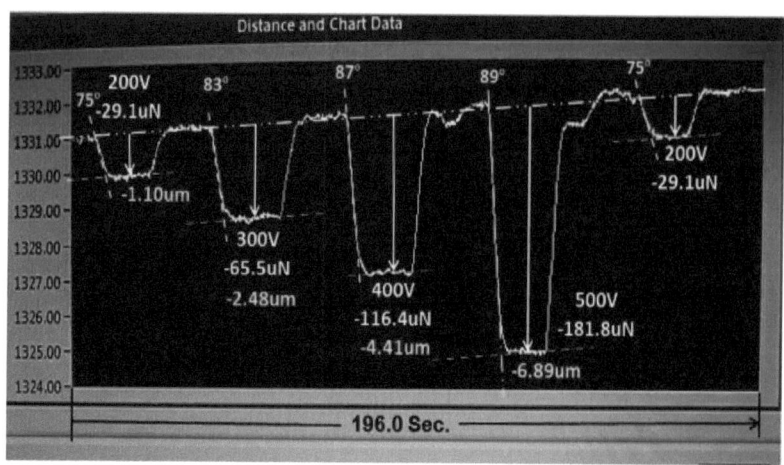

Figure 42: Integrated copper "Frustrum" (9.3 kg) force calibration pulses. (The uN stands for micronewtons.)

the Earth's magnetic field and give a Lorentz force. All the wires I have on the cavity are either twisted pair or twisted with shielding. So most of that area was reduced to a very small value, but I still ended up having RF ground loops going through the structure to ground. The only way I could really account for that was to do a dummy test, for a given power and DC current and evaluate the offset.

Williams: When the cavity heats up, how much can the cavity change in length?

March: The cavity can change by as much as 10-20 μm, both in length and diameter, which is enough to change the resonant frequency by 1-2 MHz. This is significant when the loaded Q-factor gives a bandwidth of 300 kHz.

I was never happy with the calibration of baseline noise of the vibration environment. I changed the magnetic damper design by enclosing the magnetic damper in a 1/4 inch thick iron tube, to shield any nearby wires from its magnetic field. See Fig. 43 & 44. I added one more neodymium magnet and substituted a copper metal plate for the aluminum plate used previously. That greatly enhanced the damping so that I almost got rid of the baseline noise. See Fig 45. (For comparison see Figs. 11 & 12 for the first generation damper.)

Someone suggested that a small (center of mass) COM tilt along the X axis of the balance arm would explain the apparent drift in the baseline

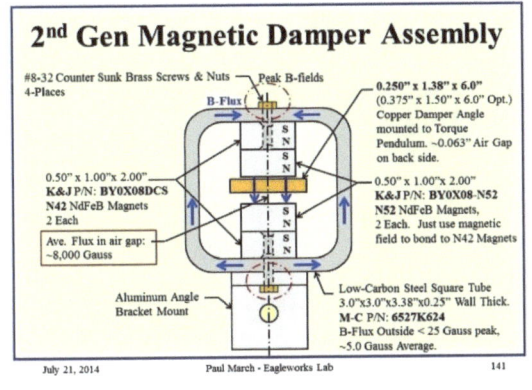

Figure 43: The 2nd generation magnetic damper schematic drawing.

Figure 44: The 2nd generation magnetic damper encased in a 1/4 inch thick iron tube, for shielding.

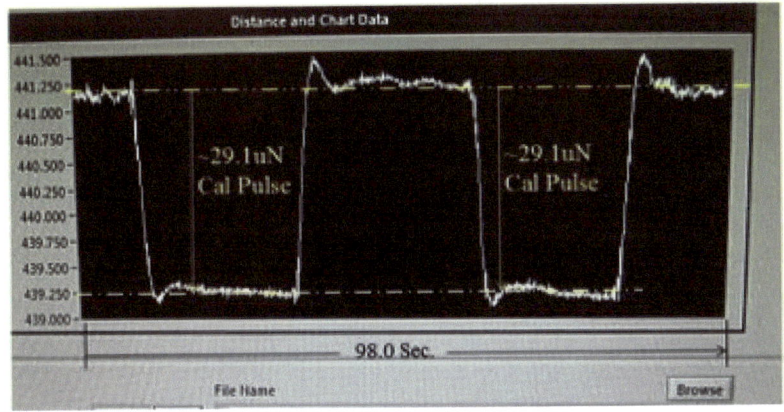

Figure 45: The 2nd generation magnetic damper wth 1.38 x 0.25 inch copper blade, using a 200 V calibration pulses. (The uN stands for micronewton.)

seen in some of the thrust waveforms. When the cavity has the orientation shown below its COM shifts to the left. This would reduce the tilt, resulting in an increase in brightness of the reflected light the linear displacement sensor measures; due to the mirror position being closer to an optimal perpendicular position with respect to the light beam. The increase in brightness corresponds to a decrease in distance; hence the negative slope. With the device mounted the other way the shift in COM increases the tilt. This reduces the reflected light and is registered as an increase in distance. No actual motion of the beam occurs. This apparent motion is an optical artifact. This assumes the Philtec linear displacement sensor is used on the far side. If it is used on the near side a

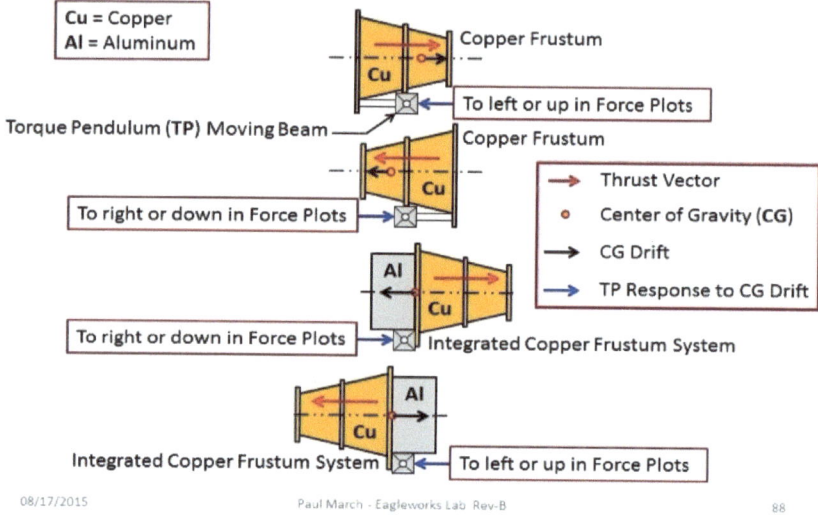

Figure 46: Torque pendulum's Force Baseline Drift due to lateral mass offset. Thermally driven center of gravity drifts.

small counter-clockwise tilt along the X axis of the beam would produce the same effect; except requiring much less rotation from the change in COM. See Fig. 47. There is a tilt in the torque pendulum's (TP) axis of

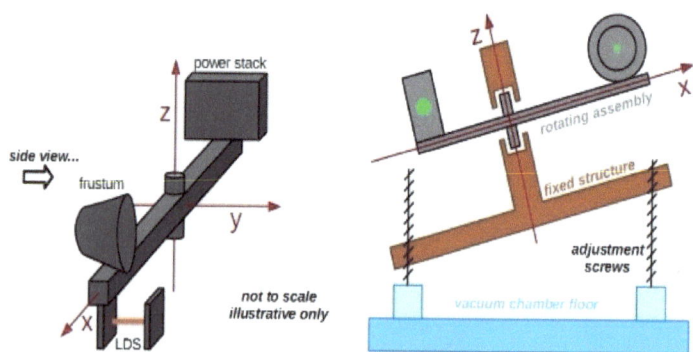

Figure 47: Torque pendulum's, very exaggerated tilt. Diagram care of "Frobnicat" from NASA Spaceflight forum.

rotation with the RF power supply lower than the test article, as shown in Fig. 47. Using my 24 inch long level, I use about a 1/4 bubble of tilt. From that point on its a matter of fine tuning the TP response by adding or removing small weights around the test article and/or adjusting the tilt angle with a micrometer adjusting the length of the TP support under the test article.

As the cavity expands you get center-of-gravity shifts which can torque the pendulum and cause it to move. So we had to spend quite some time quantifying exactly how much the pendulum moves due to small shifts in the center of gravity of the EM-drive. See Fig 46. Then we had to separate that out from the force calculations. We went into the details in our Dec 2016 paper so I won't dwell on the details here [18].

4. Conclusions

During this Sept. 2016 Estes Park Advanced Propulsion Workshop, I had the opportunity to exchange ideas and information on propellantless propulsion (P-P) and its current state of maturity. We reviewed various ways to build P-P effect devices based on either the Mach-Effect or quantum-vacuum conjectures that both rely on the cosmological gravitation field to convey momentum to and from the P-P thrusters in question.

The Mach-Effect Gravity Assist (MEGA) P-P thrusters can utilize both low frequency (35 kHz) driven, high permittivity ($e - r = 1500$), piezoelectric and electrostrictive dielectric capacitor discs used in vibrating stacks to generate these P-P effects. We also examined P-P EM-drive like thrusters that utilized microwave frequencies ($\sim$2.0 GHz) driven asymmetrical (frustum) resonant cavities, named by its inventor Roger Shawyer. These EM-drive microwave thrusters utilized either amorphous low-permittivity ($e - r = 2.3$), but high electrostrictive dielectrics like polyethylene (P-E) or Teflon for their active media, or just the copper metal down to 5× its AC skin depth, that makes up these frustum resonant cavities as the active P-P agent. Both approaches are currently able to produce in-vacuum vetted thrust levels in the micro-Newton to tens or even low hundreds of micro-Newtons, at-this-time. Ways to increase these current thrust levels into the milli-Newton level and above were reviewed and discussed.

In regards to the theory of operation, we primarily talked about Jim Woodward's M-E conjecture and how that can be applied to both his Mach-Effect Gravity Assist (MEGA) drives and Roger Shawyer's microwave powered EM-drives. When Woodward disclosed that three other labs outside the USA had replicated the testing of his PZT stack based MEGA drives in their own facilities, it dawned on me that the P-P thruster community may have finally reached their "Chigago Pile Moment" where enough experimental data validating the P-P data in question comes together to form a body of knowledge that can then be used to build bigger and better P-P thrusters needed to build vehicles like IXS Clarke solar system cruiser [1] displayed at the front of this presentation. As to what is the best way to build these P-P drives has yet to be determined and may take many roads forward, but there is now at least two ways to build viable P-P thrusters that can be utilized in deep space transport design [8].

References

[1] Mark Rademaker, http://mark-rademaker.blogspot.com/

[2] David Brady, Harold White, Paul March, James Lawrence, and Frank Davies. "Anomalous Thrust Production from an RF Test Device Measured on a Low-Thrust Torsion Pendulum", 50th AIAA /ASME /SAE /ASEE Joint Propulsion Conference, Propulsion and Energy Forum, July 28-30, 2014, Cleveland, OH (AIAA 2014-4029) http://dx.doi.org/10.2514/6.2014-4029, http://arc.aiaa.org/doi/abs/10.2514/6.2014-4029

[3] H. White and P. March, "Advanced Propulsion Physics: Harnessing the Quantum Vacuum", 2015
http://www.lpi.usra.edu/meetings/nets2012/pdf/3082.pdf

[4] Roger Shawyer, see the EM-drive website at http://www.emdrive.com.

[5] J. F. Woodward, *Making Starships and Stargates*, (Springer press Dec 2012).

[6] Guido Fetta, "Numerical and Experimental Results for a Novel Propulsion Technology Requiring no On-Board Propellant", http://arc.aiaa.org/doi/abs/10.2514/6.2014-3853, presented at the 50th AIAA /ASME /SAE /ASEE Joint Propulsion Conference, Propulsion and Energy Forum, July 28-30, Cleveland OH (2014). http://cannae.com.

[7] Roger Shawyer, "Second generation EM-drive propulsion applied to SSTO launcher and interstellar probe". Acta Astronautica **116**, 166-174 (2015). This paper has a YCBO superconducting cavity in it. www.emdrive.com/IAC14publishedpaper.pdf

[8] Harold "Sonny" White, Warp drive space ships of the future: https://www.youtube.com/watch?v=4zP0xm6r_A

[9] Roger Shawyer, "A Theory of Microwave Propulsion for Spacecraft",
http://www.rexresearch.com/shawyer/shawyer.htm\#gb233, New Scientist 2572, p24, 7 October (2006).

[10] Allen H. Yan, B. C. Appel & J. G. Gedrimas, "MilliNewton Thrust Stand Calibration Using Electrostatic fins", 47th AIAA Aerospace Sciences Meeting, Including the New Horizons Forum and Aerospace Exposition, 5th-8th Jan. Orlando FL (2009). AIAA2009-212.

[11] Greg Egan, "Resonant Modes of a Conical Cavity", http://www.gregegan.net/SCIENCE/Cavity/Cavity.html

[12] Frank Davies, did a preliminary analysis of the truncated cone cavity using COMSOL. He supplied pictures of the electromagnetic modes from 900 MHz to 2.5 GHz. These were later verified experimentally.

[13] COMSOL Multi-physics is a finite-element general-purpose software package, for solving and modeling various engineering and physics-based problems, especially coupled-problems. http://www.comsol.com.

[14] Yang, Juan et al., "Thrust Measurement of an Independent Microwave Thruster Propulsion Device with Three-Wire Torsion Pendulum Thrust Measurement System", Journal of Propulsion Technology, Vol.37, (2) Feb. 2016.

[15] College students conducted an experiment on a cylinder with a HDPE disc at one end. They were having trouble with their setup and their results were inconclusive. California Polytechnic State University, San Luis Obispo.
https://www.linkedin.com/in/kurtwadezeller

[16] c-flex flexural bearings, https://c-flex.com/ D-30 model has a load capacity of 100 pounds.

[17] Silicone spray used to prevent oxidation of the copper,
http://www.tri-m.com/products/trim/
spec-sheets/422B _TDS_Silicon.pdf

[18] H. White et al. "Measurement of Impulsive Thrust from a Closed Radio-Frequency Cavity in Vacuum", AIAA Journal of Propulsion & Power (JPP) Nov. 17th 2016. http://arc.aiaa.org/doi/full/10.2514/1.B36120

Revolutionary Propulsion Research at TU Dresden

M. Tajmar
Institute of Aerospace Engineering
Technische Universität Dresden
01062 Dresden, Germany

Since 2012, a dedicated breakthrough propulsion physics group was founded at the Institute of Aerospace Engineering at TU Dresden to investigate revolutionary propulsion. Most of these schemes that have been proposed rely on modifying the inertial mass, which in turn could lead to a new propellantless propulsion method. Here, we summarize our recent efforts targeting four areas which may provide such a mass modification/propellantless propulsion option: Asymmetric charges, Weber electrodynamics, Mach's principle, and asymmetric cavities. The present status is outlined as well as next steps that are necessary to further advance each area.

1. Introduction

Present-day propulsion enables robotic exploration of our solar system and manned missions limited to the Earth-Moon distance. With political will and enough resources, there is no doubt that we can develop propulsion technologies that will enable the manned exploration of our solar system.

Unfortunately, present physical limitations and available natural resources do in fact limit human exploration to just that scale. Interstellar travel, even to the next star system Alpha Centauri, is some 4.3 light-years away which is presently inaccessible – on the scale of a human lifetime. For example, one of the fastest manmade objects ever made is the Voyager 1 spacecraft that is presently traveling at a velocity of 0.006% of the speed of light [1]. It will take some 75,000 years for the spacecraft to reach Alpha Centauri.

Although not physically impossible, all interstellar propulsion options are rather mathematical exercises than concepts that could be put into reality in a straightforward manner. For example, from all feasible propulsion systems ever proposed the highest performance is expected from nuclear bombs which are detonated behind the spacecraft (this concept was originally developed under the name Project Orion) [2]. Even such a system would require an order of magnitude more warheads than presently available just to achieve a fly-by mission to our nearest star within a human lifetime.

Even if we could achieve a good fraction of the speed of light, our practical action radius for human-return missions would still be limited to about 10 light-years which includes a maximum of 10 stars around us where no planets have been detected so far. According to the "Maccone Distribution" [3], the next civilization would be most probably some 2000 light-years away which would be inaccessible even with hypothetical light-speed propulsion systems. It is quite clear that we need some sort of breakthrough in propulsion physics to circumvent these limits and enable practical – and affordable – human exploration well beyond our solar system.

Following the spirit of past programs such as NASA's breakthrough propulsion physics and BAE Systems Project Greenglow, we started our own breakthrough propulsion physics program [4] investigating:

1. Theory: Explore theoretical concepts that can lead to a practical Space/Warp drive, new approach to gravity that can be experimentally tested, etc.

2. Mass Modification: Investigate experimentally if mass is influenced by temperature, rotation, charge/polarization, etc.

3. New Gravitational-Like Fields: Carry out experiments to investigate if gravitational/ frame-dragging fields can be enhanced in the lab e.g. by strong discharges through superconductors

4. Testing other Claims: Critically assess claims by others on revolutionary propulsion concepts of new physical effects that may lead to a breakthrough in propulsion and/or power.

Recent work by our group include a critical evaluation of the EMDrive [5], a replication of the Wallace gravitational generator [6], a superconducting gravitational impulse generator [7], [8], the evaluation of error sources when testing weight changes of mechanical gyroscopes [9], an evaluation of the claimed electrostatic torque effect [10] as well as a possible space drive concept [11], [12] and theoretical work on a connection between electromagnetism, mass and quantum theory [13].

As classical propulsion (force and Tsiolkovsky rocket equation, etc.) is based on Newton's mechanics, which in turn relies on inertia, it is quite straightforward to think that any new type of propulsion will probably involve a change in the inertial mass. Two main approaches have appeared so far:

1. Negative mass: If we find or create a substance with negative inertial mass, put it next to a normal positive inertial mass and allow for a force between them (e.g. by charging them up with opposite polarity), this so-called gravitational dipole will start to

self-accelerate. That is a consequence of Newton's mechanics extended to negative inertia, which does not violate energy or momentum conservation as negative inertia also represents negative energy/momentum. The self-accelerating system therefore produces no net energy/momentum itself. This concept was first proposed by Forward [14] and recently even experimentally verified in an optical analog experiment with self-accelerating photons [15].

2. Variable/Oscillating mass: It may not be necessary for a revolutionary propulsion device to have negative inertial mass, it could be sufficient to have an inertial mass that is oscillating. If we imagine such a mass that we push when it is heavy and pull back when it is lighter, such a system could indeed produce a net momentum without spending propellant. As recently explicitly shown by Wanser [16], momentum conservation does only apply to a system with constant mass. Our oscillating mass system clearly violates this condition providing a method of producing real propellant-less thrust. Of course energy must be spent in order to modify mass and to push/pull it back and forth. Properly written down, also this approach does not violate any physical conservation principle.

Of course, the real challenge here is to produce macroscopic quantities of negative or oscillating inertial mass. So far, the properties of negative inertial mass have been mimicked in experiments using effective mass inside certain boundaries only (e.g. neutrons inside a crystal [17], or photons inside fibers [15]). How shall real negative mass exist outside such special boundaries? Oscillating inertial masses are much simpler to imagine. For example, charging and discharging a capacitor will change its mass by simply following $E = mc^2$. Unfortunately, c^2 is a large number so the resulting mass fluctuation will be very small. Of course the availability of high-frequency technology up to the THz range may compensate some of that if properly done.

The approach currently pursued at TU Dresden is to investigate four different possibilities to achieve negative/oscillating inertial mass as shown in Fig. 1. This paper will give an overview of the present status for each of the research lines.

2. Asymmetric Charges

According to Einstein's famous equation $E = mc^2$, all non-gravitational sources of energy contribute to mass (the energy of the gravitational field cannot be localized according to the equivalence principle [18]). Boyer [19] showed that two opposite charges should lose weight as the electrostatic potential energy between dissimilar charges is always negative. Considering two charges, the energy of the whole system is given as:

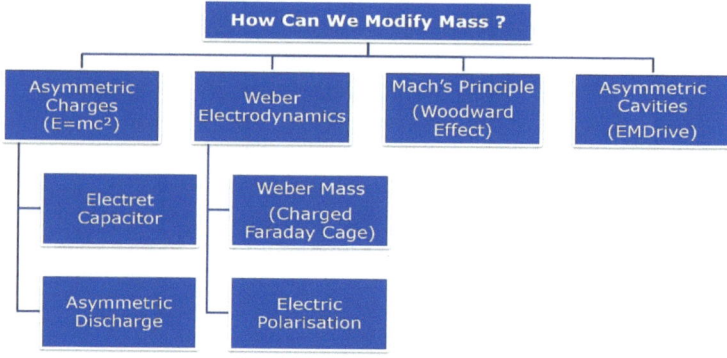

Figure 1: Mass Modification Approach

$$U = m_1 c^2 + m_2 c^2 + \frac{1}{4\pi\epsilon_0} \frac{q_1 q_2}{r} \quad (1)$$

where r is the separation distance between the charges, and m and q is the respective mass and amount of charge. It is now straightforward to see that if the two charges have opposite signs, the electrostatic potential energy is reducing the total mass of the system by

$$\Delta m = \frac{1}{4\pi\epsilon_0} \frac{q_1 q_2}{r c^2} \quad (2)$$

Of course the main question here is where this change in mass is actually localized. Is the delta-mass equally split between the charges involved, or is this delta mass only visible for the system of charges as a whole? If the actual mass of a charge would be modified, then this could open up the possibility to use this effect for our propellantless-propulsion scheme.

The contribution of electrostatic energy to mass is actually a century-old question. The simplest configuration is the one of a single electron acting on itself (self-energy). Initially J.J. Thompson derived the so-called electromagnetic mass (1881) and at the beginning of the 20th century it was thought that this electromagnetic contribution actually is responsible the whole mass of the electron. That changed of course with the development of relativity and quantum theory. Still, self-energy contributions to mass and the resulting perturbations to the classical motion of particles is an active field of research (e.g. [20]). However, self-energy contributions and contributions to each charge from multiple charge interactions are very different scenarios.

Brillouin [21] studied this question and argued that as almost all energy associated with the electric field is localized within the classical electron radius, the mass change should localize at the individual particles as well. If we consider a point-particle with charge Q, the energy of the electric field from infinity towards a radius R is defined as

$$U = \frac{1}{8\pi\epsilon_0} \frac{Q^2}{R} \qquad (3)$$

Accordingly, the mass associated with that energy can be expressed as

$$M_q = \frac{U}{c^2} = \frac{Q^2}{8\pi\epsilon_0 c^2}\frac{1}{R} = \frac{Q\phi}{2c^2} \qquad (4)$$

where ϕ is the electric potential. From these equations, it's clear that the mass diverges as R approaches zero and therefore a finite radius is required for the charged particle. That was how the classical electron radius was introduced. Still, the introduction of an arbitrary radius to justify that the energy of the field materializes as a mass change for every particle involved is not fully convincing.

Contrary to this classical approach that summarizes the energy from infinity towards an arbitrary radius (outside view), a more modern approach is given by the Reissner-Nordström metric which describes the field equations of a mass M with charge Q as

$$\begin{aligned}ds^2 &= \left(1 - \frac{2GM}{rc^2} + \frac{GQ^2}{4\pi\epsilon_0 r^2 c^4}\right) c^2 dt^2 - \left(1 - \frac{2GM}{rc^2} + \frac{GQ^2}{4\pi\epsilon_0 r^2 c^4}\right)^{-1} dr^2 \\ &\quad - r^2\left(d\theta^2 + \sin^2\theta d\phi^2\right)\end{aligned}$$

where the line element is approximated using

$$g_{00} \cong 1 + 2\frac{\Delta U}{c^2}, \quad \Delta U = -\frac{GM}{r} + \frac{GM_q}{r} = -\frac{GM}{r} + \frac{GQ^2}{8\pi\epsilon_0 c^2}\frac{1}{r^2} \qquad (5)$$

This can be considered the length element inside the mass, as here we are really dealing with the equations of motion of the charged mass itself. As the gravitational potential energy is negative (mass attracts mass) but the electrostatic potential energy is positive, this charge-energy correction here acts as a negative mass component (Reissner-Nordström repulsion) at $r < R$, where R is the event horizon. We see that this is exactly opposite to the electromagnetic mass view that assigned a positive mass due to the electrostatic self-energy of the field. Bringing both views together, one may even say that indeed the motion of charged particles will be affected by the electrostatic field, but there is no net mass gain at the location of the particle as the negative contribution there is balanced out by the positive contribution due to the energy density of the field towards infinity.

Still, this question has not been experimentally assessed thoroughly. The contribution of an electrostatic potential to mass (electrostatic redshift) was experimentally investigated with a null result by Kennedy et al and Drill in the 1930s [22], [23]. Woodward and Crowley [24] pointed out that this result was to be expected using the Reissner-Nordström metric to predict the effect and the instrumentation resolution at that time. New experiments will be necessary to probe such an effect.

On the promise that electrostatic fields may influence a particle's rest mass, we recently published a configuration called the electret capacitor which could enable the utilization of this effect for propulsion purposes [11]. A capacitor typically consists of two sheets of metal with a dielectric in between. If the capacitor is charged, a certain amount of charge leaves one surface to go to the second one. Therefore, the charge density on both plates is equal but with different polarities. For the electret capacitor, two electrets (sheets of dielectrics with permanent electric charges on them) with different charge densities are opposite to each other, creating a new electrostatic situation where the positive self-energy from the interaction of charges with the same polarity can be outbalanced by the negative interaction energy between the charges with different polarities. In certain geometrical and charge density configurations, a negative energy larger than the positive rest mass energy of the charges from one side of this electret capacitor may be created, which could be used as a negative inertial mass source for propellantless propulsion.

Apart from the electric configuration, discharges in a highly asymmetric electric field may also provide the necessary boundary for charges to behave as negative inertial masses which may result in a novel propulsion scheme.

3. Weber Electrodynamics

In parallel to the development of Maxwell's equations, Wilhelm Weber proposed a force that also covered all known aspects of electromagnetism (Ampere, Coulomb, Faraday and Gauss's laws) and incorporated Newton's third law in the strong form, that is that the force is always along the straight line joining two charges [25] (which also implies the conservation of linear and angular momentum). However, Weber's electrodynamics also gives rise to new effects such as the change of the effective inertial mass of a charge inside a charged spherical shell which we could exploit for negative matter propulsion. Assis proposed an extension to Weber's electrodynamics that allows the derivation of a gravitation-type force [26], [27]. This extended model may be used to actually modify mass itself. Here we will give a short overview of both approaches.

A. Weber Mass (Charged Faraday Cage)

Weber's force expression and the related potential energy is given by

$$\mathbf{F} = \frac{q_1 q_2}{4\pi\epsilon_0} \frac{\hat{\mathbf{r}}}{r^2} \left(1 - \frac{\dot{r}^2}{2c^2} + \frac{r\ddot{r}}{c^2}\right), \quad U = \frac{q_1 q_2}{4\pi\epsilon_0} \frac{1}{r}\left(1 - \frac{\dot{r}^2}{2c^2}\right), \quad (6)$$

where q_1 and q_2 are the respective charges and r is the distance between them. If we now consider a single charge inside a charged spherical dielectric shell (in order to ignore eddy currents or mirror charges), we must integrate the force and sum up all the interaction between the single charge inside the shell and all other charges along the shell. Surprisingly, a net force remains that acts on the single charge when it accelerates inside the shell [28] given by

$$\mathbf{F} = \frac{qQ}{12\pi\epsilon_0 c^2 R} \cdot \mathbf{a} = \frac{q\phi}{3c^2} \cdot \mathbf{a}, \quad (7)$$

where Q is the charge on the shell, R the shell's radius and ϕ the electrostatic potential inside the shell. Classically, no force is expected on a charge inside a charged shell as the electric potential is constant and therefore no electric and no force acts on charges inside. According to Weber's electrodynamics, this force is proportional to acceleration of the charge and therefore influences the charge's inertial mass. If the total inertial mass is now the sum of the unaffected mass and the Weber mass, we may express the effective mass of the charge as

$$m^* = m - \frac{qQ}{12\pi\epsilon_0 c^2 R} = m - \frac{q\phi}{3c^2} \quad (8)$$

The equation predicts that a change in mass should be quite observable in a dedicated laboratory experiment. Considering a dielectric shell with a radius of 0.5 m charged up to 1.5 MV, we could expect to double an electron's mass – or reduce it to zero depending on the shell's charge polarity. In fact, up to a numerical factor, that result is very close to the one for the electromagnetic mass (see Eq. (4)).

Mikhailov published a number of experiments where such an effect was indeed observed. First, he put a neon glow lamp inside a glass shell that was coated by a thin layer of GaIn and an RC-oscillator inside a Faraday shield below [29]. The coated glass shell imitates the charged dielectric shell as originally proposed by Assis. Mikhailov assumed that the frequency of the lamp is directly proportional to the electron's mass. Indeed, he observed that the lamp's frequency changed if he charged the sphere as predicted by Equ. (9) within a factor 3/2. In a second experiment, the neon lamp was replaced by a Barkhausen-Kurz generator leading to similar results [30]. Finally, the neon-lamp experiment was repeated with two charged concentric shells showing that the frequency/mass effect from charging up the first shell can be counterbalanced by oppositely charging the outer shell [31].

Junginger and Popovich [32] repeated the neon glow lamp experiment and implemented an optical counter instead of electrically measuring the frequency of the lamp – and observed a null result. Also Little et al [33] performed a similar replication and observed a null result with optical counters and observed that the electric measurement of the lamp's frequency may be influenced by the Faraday's shield potential depending on the coupling capacitor used (however the signature of the effect was a parabola instead of the linear relationship as obtained by Mikhailov). At TU Dresden, we tried to replicate Mikhailov's setup and implemented an optical counter in parallel. Indeed, we could also verify the variation that Mikhailov has seen and traced it back to influence of the coupling capacitor. Running the experiment with an optical counter also produced a null effect.

However, we then asked ourselves how representative a neon discharge is with respect to the single electron prediction from Weber/Assis. A plasma discharge produces a significant current and a number of ions in close proximity to the electrons. This setup may therefore not be representative at all in order to test this prediction. Mikhailov's second setup used a Barkhausen-Kurz generator where an electron cloud is oscillating around a grid with high frequency. This frequency f should be closely linked to the mass of the electron as given by:

$$f \approx \sqrt{\frac{e\phi}{2m}} \cdot \frac{1}{\ell} \qquad (9)$$

where ℓ is the distance from the cathode to the anode. Mikhailov did not measure the frequency directly in his setup but only qualitatively. We decided to make a replication using both the same tube as well as others that are known to produce Barkhausen-Kurz oscillations. We then put the tube inside a 3D printed shell with a metallic layer that could be biased. Using an Advantest R3261A signal analyzer, the actual frequency of the tube during biasing the spherical shell could be monitored as shown in Fig. 2.

The following observations were made (a detailed description of the experiment will be presented elsewhere):

- The original Mikhailov setup did not produce Barkhausen-type oscillations as the frequency did not scale with the square-root of the applied voltage to the grid.

- We replaced the tube and electronics successfully to observe Barkhausen-type oscillations with the correct characteristics.

- The frequency of the maximum signal peak emitted signal was tracked while varying the potential applied to the metallic sphere.

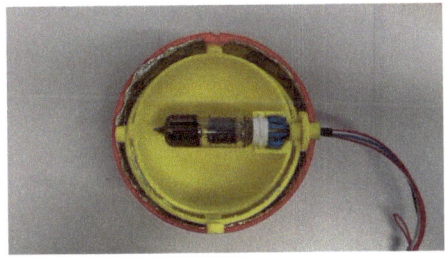
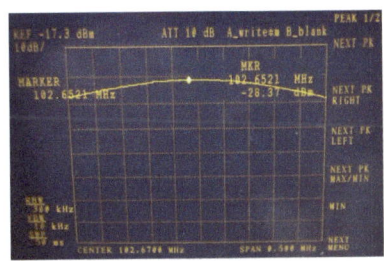

(a) Inside Charged Sphere　　　　(b) Frequency Measurement

Figure 2: Barkhausen-Kurz Generator Setup.

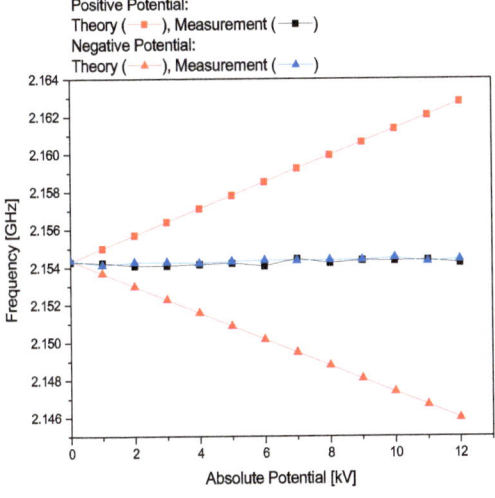

Figure 3: Observed Frequency Variation of Maximum Signal Peak (Average of Three Test Runs) with Respect to the Expected Variation according to Weber/Assis

The result is shown in Fig. 3. As it can be seen, our resolution was more than an order of magnitude better to see the predicted effect but no variation with the applied potential could be seen. However, it must be noted that the width of the signal was about 5 MHz which is in the range of the expected variation (8 MHz at 12 kV).

Of course, also here we have to ask if the experimental setup correctly represents the case predicted by Weber/Assis. For example, here we have an electron cloud instead of a single electron and the approximation of the Barkhausen oscillation in Equ. (10) also leaves room for correction factors that could possibly change our expected variation. Further experiments with different setups are necessary to look for an electrostatic influence on mass to find a definite answer.

B. Electric Polarization

Assis [26], [27] proposed an extension to Weber's electrodynamics that allowed him to derive gravitational and inertial-type forces from electrodynamics. His model is based on two assumptions:

1. Mass is composed of two opposite charges that vibrate with a certain amplitude and frequency. This can be considered a string-type approach.

2. Weber's potential Equ. (7) is actually a first order approximation valid for Maxwellian electromagnetism. Assis generalizes this equation with high-order terms as follows:

$$U = \frac{q_1 q_2}{4\pi\epsilon_0 r}\left(1 - \alpha\left(\frac{\dot{r}}{c}\right)^2 - \beta\left(\frac{\dot{r}}{c}\right)^4 - \gamma\left(\frac{\dot{r}}{c}\right)^6 \cdots\right), \quad (10)$$

where α is known as 0.5 and the other coefficients are assumed on the order of unity without knowing their precise value. Then he calculates the force between two oscillating dipoles with charge q, amplitude A and angular frequency ω by averaging over time and the three possible orientations (x, y, z) of the oscillating strings to arrive at

$$F = -\frac{7\beta}{18}\left(\frac{q_{1+}q_{2+}}{4\pi\epsilon_0 r^2}\right)\frac{A_{1-}^2\omega_1^2 A_{2-}^2\omega_2^2}{c^4}\left(1 + \frac{\gamma}{\beta}\frac{45\dot{r}^2 - 18r\ddot{r}}{7c^2}\right) \quad (11)$$

This looks like an always attractive force between the oscillators comparable with a similar $1/r^2$ dependence like gravity. The second-order correction term in the equation is identified with inertia. Of course, there are a number of free parameters (q, A, ω and the coefficients β and γ) that make it difficult to predict actual masses. However, recently we could show that this model allows the correct prediction of the maximum possible point mass which is equal to the Planck mass allowing to derive Planck's constant and the fine-structure constant with only one free coefficient [13]:

$$\hbar = \frac{h}{2\pi} = \frac{7\pi^3 e^2 \beta}{72 c \epsilon_0} = 2.92 \times 10^{-35}\beta \quad (12)$$

which matches the known value exactly for $\beta = 3.62$ (it is on the order of unity as Assis assumed). This is a remarkable result as it is the first derivation of the core assumption of quantum theory from an electromagnetic and gravitational model, providing a possible link between these cornerstones of modern physics and possibly an alternative to the Higgs model approach to explain mass.

If the Assis mass model is correct, then it may be possible to influence mass, e.g., due to electric polarization which is then influencing the orientation of the oscillating dipoles and therefore the average force between them. Apart from theoretical models to study such scenarios, we are currently testing the influence of highly polarized wax-electrets on their weight as a function of polarization and time. Similar tests were recently reported in a patent from Kita [34] where he claimed changes as high as 140 mg for samples with a weight of 278 g. We started our own wax-based electret production (45% carnauba wax, 45% resin and 10% bee wax) that were electrically polarized inside a capacitor with up to 10 kV during their cooling down phase. We used glass containers in order to limit any gas exchange with the environment which turned out to be very critical. That limited the observed weight changes in our experiments for samples with up to 200 g (including the container) to a few milli-grams only (see Fig. 4) [35]. We are presently further improving the setup in order to trace temperature and humidity changes in order to find an explanation for the observed drifts. Then we will proceed with measurements of different type of electrets or capacitors in order to investigate this mass change possibility.

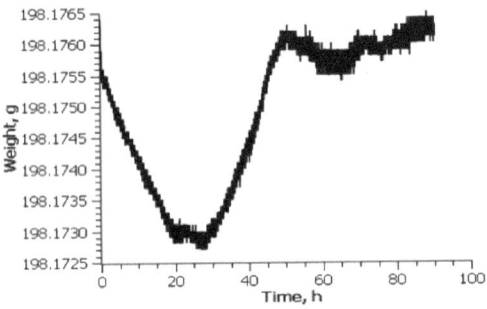

Figure 4: Weight Change of Polarized Electrets over Time [35]

4. Mach's Principle (Woodward Effect)

Mach's principle is a concept in physics that tries to explain inertia [36]. It had been a guiding principle for A. Einstein in the development of his general relativity theory. Although there are many different interpretations, a simple explanation would be: "mass out there influences inertia here". It means that every mass is connected to all the masses of the whole universe by gravitational forces, which in turn is the cause for inertia. Some consequences of Einstein's theory can be indeed viewed as

Machian, like the dragging of space-time by rotating objects which then influences objects in their close vicinity.

Over many years, J.F. Woodward used Mach's principle to propose a scheme that he calls transient mass fluctuations [37], which suggests that measureable changes in the inertial mass of a body can be created due to high-frequency oscillations which are caused by a back-reaction of the universe on the oscillating test mass. His derivation is based on a flat-space, low-velocity relativistic evaluation of the four-divergence of the back-reaction field that arises from the gravitation of the universe. Here we will present a simple analysis using linearized general relativity theory that arrives at similar conclusions without any necessary assumptions.

Linearizing general relativity is an approximation scheme valid for test masses at slow velocities (with respect to the speed of light), in an environment that is not dominated by large gravitational fields (e.g. black holes), which is a good representation of our laboratory boundaries. The starting point is the Einstein field equation, where the metric tensor $g_{\mu\nu}$ is treated as flat spacetime $\eta_{\mu\nu}$ with a perturbation component $h_{\mu\nu}$:

$$R_{\mu\nu} - \frac{1}{2}g_{\mu\nu}R = \frac{8\pi G}{c^4}T_{\mu\nu}, \quad g_{\mu\nu} \cong \eta_{\mu\nu} + h_{\mu\nu}. \tag{13}$$

By using the definitions

$$\bar{h}_{\mu\nu} = h_{\mu\nu} - \frac{1}{2}\eta_{\mu\nu}h, \quad \bar{h}_{00} = \frac{4\phi_g}{c^2}, \quad T_{00} = \rho c^2, \tag{14}$$

it is possible to simplify Einstein's equation to

$$\frac{1}{c^2}\frac{\partial^2}{\partial t^2}\bar{h}_{\mu\nu} - \nabla^2 \bar{h}_{\mu\nu} = -\frac{16\pi G}{c^4}T_{\mu\nu} \tag{15}$$

Now, one usually takes as a first order only static solutions, which ignores the first term on the left side, that immediately leads to Newton's gravitational force law:

$$\nabla \cdot \mathbf{g} = -\nabla^2\phi_g = -\frac{4\pi G}{c^2}T_{00} = -4\pi G\rho_0 \tag{16}$$

where $\mathbf{g} = -\nabla\phi_g$ is the gravitational force per unit mass. As we are looking for transient solutions, we will now relax the approximation for the static solution and keep the first term in Equ. (15). This then leads to a deviation from Newton's law that is given as:

$$\frac{1}{c^2}\frac{\partial^2 \phi_g}{\partial t^2} - \nabla^2\phi_g = -\frac{4\pi G}{c^2}T_{00} = -4\pi G\rho_0 \tag{17}$$

and therefore

$$\begin{aligned}\nabla \cdot g &= -\nabla^2 \phi_g = -4\pi G \rho - \frac{1}{c^2}\frac{\partial^2 \phi_g}{\partial t^2} \\ &= -4\pi G \left(\rho_0 + \frac{1}{4\pi G c^2}\frac{\partial^2 \phi_g}{\partial t^2}\right)\end{aligned} \quad (18)$$

By comparing Eqs. (16) and (18), we see that time-varying terms lead to a change in the body's density (or mass by integration over its volume) that is independent of the gravitational constant G, which make such terms very large compared to "static" density (mass). This structure looks similar to displacement currents in Maxwell's equations. In the introduction, we discussed the example of a capacitor that is being charged and discharged and therefore varies its mass due to $E = mc^2$, making the mass changes too small to be observed. However, Eq. (18) tells us that fast mass changes are coupling much stronger to the gravitational field (by the factor $1/G \approx 1.5 \times 10^{10}$) than static masses do, which should make this effect indeed observable.

The change in density can be expressed as

$$\begin{aligned}\delta\rho_0 &= \frac{1}{4\pi G c^2}\frac{\partial^2 \phi_g}{\partial t^2} = -\frac{\phi_g}{4\pi G c^2 m_0}\frac{\partial^2 m_0}{\partial t^2} \\ &= -\frac{\phi_g}{4\pi G c^2 \rho_0}\frac{\partial^2 \rho_0}{\partial t^2} = \frac{1}{4\pi G \rho_0}\frac{\partial^2 \rho_0}{\partial t^2}\end{aligned} \quad (19)$$

where we used $\phi_g = -Gm_0/r$ for the gravitational potential and $\phi_g/c^2 = -1$ which was derived by Sciama [38] due to the interaction of the gravitational potential throughout the whole universe, which is of course the concept of Mach's principle. This equation is similar to the one from Woodward (first order term) and clearly shows that indeed transient Mach-type fluctuations are predicted by general relativity theory without the introduction of new physics.

So far, over the years many tests have been published by Woodward's lab [37,39,40] and others [41,42]. His design is based on piezo crystals that act both as capacitors that trigger mass changes due to rapid charging/discharging, as well as accelerators to push and pull the crystals in order to get a directional thrust as outlined in the introduction. After the implementation of a torsion balance, the observed thrusts were in the sub-μN range for the models and electronics used. Many error sources were addressed such as thermal drifts or vibration artefacts.

Still, a number of shortcomings are present that we need to tackle in order to claim an experimental effect without any doubts. Most importantly, no tests were carried out up to now with the electronics (signal generator and amplifier) on the balance in order to completely rule out interactions between them. So far, all tests used electronics outside the vacuum chamber and liquid-metal contacts that connected to the thruster on the balance. We therefore decided to build vacuum-compatible electronics that can be mounted on a thrust balance to carry out thrust measurements with a fully integrated thruster-electronics package. Our test thruster is a model that was given to us in 1999 by J. Woodward which looks similar in design, however, it contains old piezo elements with non-optimal specifications so that we expect somewhat lower thrusts compared to his present models.

Our thrust balance uses flexural bearings and is similar in its design to many other low-thrust balances with several distinct differences [43], see Fig. 5:

- Up to 25 kg of thruster and electronics weight is possible, which enables the possibility of heavy shielding if necessary.

- On-board electronics and data acquisition system with infrared wireless communication, 24 V supply through the bearings, liquid-metal contacts if needed.

- Vibration damping of the whole vacuum chamber and inside the vacuum chamber

- Calibration with electrostatic combs or voice-coil

- Use of the attocube IPS laser interferometer which enables a thrust noise down to the sub-nN regime

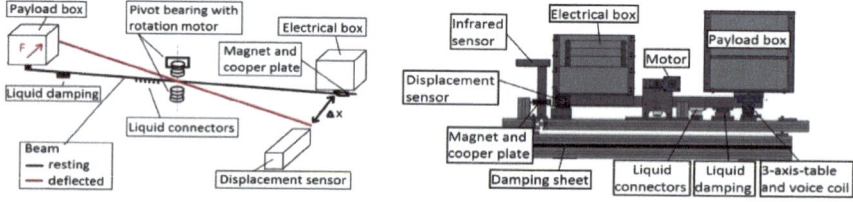

Figure 5: Thrust Balance Setup [43]

The electronics on the balance as well as the Mach-Effect thruster can be seen in Fig. 6 and the whole thrust balance inside our large vacuum chamber is shown in Fig. 7. First tests show thrust values in the sub-μN range, however, balance calibration, thermal drifts and power feeding line interactions are still under investigation before our first test campaign will be finalized.

Figure 6: Mach-Effect Thruster: Setup of Electronics and Thruster Model

Figure 7: Mach-Effect Thruster: Setup of Thrust Balance

5. Asymmetric Cavities (EM-Drive)

The EM-Drive has been proposed as a revolutionary propellantless thruster using a resonating microwave cavity [44-46]. The inventor R. Shawyer claims that it works on the difference in radiation pressure due to the geometry of its tapered resonance cavity. This may also be interpreted as a change in the effective photon mass at each side of the cavity, which somehow resembles Woodward's transient Mach-fluctuation thruster with photons instead of piezo crystals, that may ultimately lead to higher efficiencies and thrust-to-power ratios.

We attempted to replicate an EM Drive and tested it on both a knife-edge balance as well as on a torsion balance inside a vacuum chamber, similar to previous setups, in order to investigate possible side-effects through proper thermal and electromagnetic shielding. After developing a numerical model to properly design our cavity for high efficiencies in close cooperation with the EM Drive's inventor, we built a breadboard out of copper with the possibility to tune the resonance frequency in order to match the resonance frequency of the magnetron which was attached on the side of the cavity. After measuring the Q-factor of our assembly, we connected the EMDrive to a commercial 700 W microwave magnetron. An overview of the different setups can be seen in Fig. 8.

Thruster Model with Magnetron

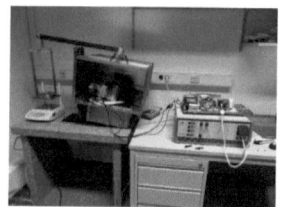
Setup with Box on Knife-Edge Balance

Setup on Thrust Balance inside Vacuum Chamber

Figure 8: EMDrive Setups

Our measurements revealed thrusts as expected from previous claims (due to a low Q factor of < 50, we observed thrusts of $\pm 20\mu N$), however also in directions that should produce no thrust. We therefore achieved a null measurement within our resolution which is on the order of the claimed thrusts. Details of the measurement can be seen found in [5].

The purpose of the test program was to investigate the EMDrive claims using improved apparatus and methods. To this end it was successful in that we identified experimental areas needing additional attention before any firm conclusions concerning the EMDrive claims could be made. Our test campaign therefore cannot confirm or refute the claims of the EMDrive but intends to independently assess possible side-effects in the measurement methods used so far. We identified the magnetic interaction of the power feeding lines going to and from the liquid metal contacts as the most important possible side-effect that is not fully characterized yet and which needs to be evaluated in the future in order to improve the resolution.

6. Conclusion

This paper summarizes the current activities towards revolutionary propulsion activities at TU Dresden. We believe this is an excellent educational topic which a great learning experience for students due to its theoretical and experimental challenges. Even an experimental null result leads to a better understanding of measurement artefacts or setup limitations which are very valuable for other similar investigations (e.g. low-thrust measurements for space thrusters). Of course, research towards totally new propulsion schemes can be very valuable to ultimately push the technological limit of our present limitations in space exploration.

Acknowledgements

I would like to thank the students involved in the activities that were summarized in this paper: Matthias Kössling (Thrust Balance and Mach Effect Thruster), Marcel Weikert (Weber Electrodynamics), Willy Stark (Mach Effect Thruster), Istvan Lörincz (Weber Electrodynamics), Tom Schreiber (Electret Measurements) and Georg Fiedler (EMDrive). Continued discussions with G. Hathaway were also greatly appreciated. In addition, I would like to thank the Estes Park organizers for their great efforts.

References

[1] R. H. Frisbee, "Advanced Propulsion for the 21st Century," J. Propuls. Power, Vol. 19, No. 6, pp. 1129-1154, 2003.

[2] M. Tajmar, Advanced Space Propulsion Systems. Vienna: Springer Vienna, 2003.

[3] C. Maccone, "The Statistical Drake Equation," in 59th International Astronautical Congress, Glasgow, Scotland, GB, Paper IAC-08-A4.1.4, 2008.

[4] M. Tajmar, "Breakthrough Propulsion Physics," 2016. [Online]. Available: http://www.tu-dresden.de/ilr/rfs/bpp . [Accessed: 09-Dec-2016].

[5] M. Tajmar and G. Fiedler, "Direct Thrust Measurements of an EMDrive and Evaluation of Possible Side-Effects," in 51st AIAA/ SAE/ ASEE Joint Propulsion Conference, 2015, AIAA 2015-4083.

[6] M. Tajmar, I. Lörincz, and C. Boy, "Replication and Experimental Characterization of the Wallace Dynamic Force Field Generator," in 51st AIAA/ SAE/ASEE Joint Propulsion Conference, 2015, AIAA 2015-4081.

[7] I. Lörincz and M. Tajmar, "Null-Results of a Superconducting Gravity-Impulse-Generator," 52nd AIAA/ SAE/ASEE Joint Propulsion Conference, AIAA 2016-4988, 2016.

[8] I. Lörincz and M. Tajmar, "Design and First Measurements of a Superconducting Gravity-Impulse-Generator," in 51st AIAA/ SAE/ ASEE Joint Propulsion Conference, 2015, AIAA 2015-4080.

[9] I. Lörincz and M. Tajmar, "Identification of error sources in high precision weight measurements of gyroscopes," Measurement, Vol. 73, pp. 453-461, 2015.

[10] D. Bojiloff and M. Tajmar, "Experimental evaluation of the claimed coulomb rotation (electrostatic torque)," J. Electrostat., Vol. 76, pp. 268-273, 2015.

[11] M. Tajmar, "Propellantless Propulsion with Negative Matter Generated by High Electrostatic Potentials," in 49th AIAA/ASME/ SAE/ASEE Joint Propulsion Conference, 2013, AIAA 2013-3913.

[12] M. Tajmar and A. K. T. Assis, "Particles with Negative Mass: Production, Properties and Applications for Nuclear Fusion and Self-Acceleration,"' J. Adv. Phys., Vol. 4, No. 1, pp. 77-82, Mar. 2015.

[13] M. Tajmar, "Derivation of the Planck and Fine-Structure Constant from Assis's Gravity Model," J. Adv. Phys., Vol. 4, No. 3, pp. 219-221, Sep. 2015.

[14] R. L. Forward, "Negative matter propulsion," J. Propuls. Power, Vol. 6, No. 1, pp. 28-37, Jan. 1990.

[15] M. Wimmer et al., "Optical diametric drive acceleration through action-reaction symmetry breaking," Nat. Phys., Vol. 9, No. 12, pp. 780-784, 2013.

[16] K. H. Wanser, "Center of mass acceleration of an isolated system of two particles with time variable masses interacting with each other via Newton's third law internal forces: Mach effect thrust," J. Sp. Explor., Vol. 2, No. 2, pp. 122-130, 2013.

[17] K. Raum, M. Koellner, A. Zeilinger, M. Arif, and R. Gähler, "Effective-mass enhanced deflection of neutrons in noninertial frames," Phys. Rev. Lett., Vol. 74, No. 15, pp. 2859-2862, 1995.

[18] C. W. Misner, K. S. Thorne, and J. A. Wheeler, Gravitation. W. H. Freeman and Comp. Ltd., 1973.

[19] T. H. Boyer, "Electrostatic potential energy leading to an inertial mass change for a system of two point charges," Am. J. Phys., Vol. 46, No. 4, p. 383, 1978.

[20] S. E. Gralla, A. I. Harte, and R. M. Wald, "Rigorous derivation of electromagnetic self-force," Phys. Rev. D, Vol. 80, No. 2, p. 24031, Jul. 2009.

[21] L. Brillouin, "The Actual Mass of Potential Energy, a Correction to Classical Relativity," Proc. Nat. Acad. Sci, Vol. 53, No. 3, p. 475, 1965.

[22] R. J. Kennedy and E. M. Thorndike, "A Search for an Electrostatic Analog to the Gravitational Red Shift Author A Search for an Electrostatic Analog to the Gravitational Red Shift," Proc. Natl. Acad. Sci. U. S. A., Vol. 17, No. 11, pp. 620-622, 1931.

[23] H. T. Drill, "A Search for an Electrostatic Analog to the Gravitational Red Shift," Phys. Rev., Vol. 56, No. 2, pp. 184-185, Jul. 1939.

[24] J. F. Woodward and R. J. Crowley, "Electrostatic Redshift," Nat. Phys. Sci., Vol. 246, No. 151, pp. 41, Nov. 1973.

[25] A. K. T. Assis, Weber's Electrodynamics. Dordrecht: Springer Netherlands, 1994.

[26] A. K. T. Assis, "Deriving gravitation from electromagnetism," Can. J. Phys., Vol. 70, No. 5, pp. 330-340, May 1992.

[27] A. K. T. Assis, "Gravitation as a Fourth Order Electromagnetic Effect,"' in Advanced Electromagnetism –Foundations, Theory and Applications, T. W. Barrett and D. Grimes, Eds. Singapore: World Scientific, 1995, pp. 314-331.

[28] A. K. T. Assis, "Changing the Inertial Mass of a Charged Particle," J. Phys. Soc. Japan, Vol. 62, No. 5, pp. 1418-1422, May 1993.

[29] V. F. Mikhailov, "The Action of an Electrostatic Potential on the Electron Mass," Ann. la Fond. Louis Broglie, Vol. 24, pp. 161-169, 1999.

[30] V. F. Mikhailov, "The Action of an Electrostatic Potential on the inertial electron mass,"' Ann. la Fond. Louis Broglie, Vol. 26, No. 1, pp. 33-38, 2001.

[31] V. F. Mikhailov, "Influence of a Field-Less Electrostatic Potential on the inertial Electron Mass," Ann. la Fond. Louis Broglie, Vol. 28, No. 2, pp. 231-236, 2003.

[32] J. E. Junginger and Z. D. Popovic, "An experimental investigation of the influence of an electrostatic potential on electron mass as predicted by Weber's force law," Can. J. Phys., Vol. 82, No. 9, pp. 731-735, 2004.

[33] S. Little, H. Puthoff, and M. Ibison, "Investigation of Weber's Electrodynamics," 2001, [Online]. Available: http://exvacuo.free.fr/div/Sciences/Dossiers/EM/Charges.

[34] R. J. Kita, "Gravitational Attenuating Material," US 8901943, 2014.

[35] T. Schreiber and M. Tajmar, "Testing the Possibility of Weight Changes in Highly-Polarized Electrets," in 52nd AIAA/SAE/ASEE Joint Propulsion Conference, 2016, AIAA 2016-4919.

[36] J. Barbour and H. Pfister, Eds., Mach's principle: from Newton's bucket to quantum gravity. Boston: Birkhäuser, 1995.

[37] J. F. Woodward, *Making Starships and Stargates*. New York, NY: Springer New York, 2013.

[38] D. W. Sciama, "On the Origin of Inertia," Mon. Not. R. Astron. Soc., Vol. 113, No. 1, pp. 34-42, Feb. 1953.

[39] H. Fearn and K. Wanser, "Experimental tests of the Mach effect thruster," J. Sp. Explor., Vol. 3, No. 3, pp. 197-205, 2014.

[40] H. Fearn and J. F. Woodward, "Experimental Null test of a Mach Effect Thruster,' arXiv, 25-Jan-2013. [Online]. Available: http://arxiv.org/abs/1301.6178.

[41] N. Buldrini, M. Tajmar, K. Marhold, and B. Seifert, "Experimental Results of the Woodward Effect on a μN Thrust Balance," in 42nd AIAA /ASME /SAE /ASEE Joint Propulsion Conference, 2006, AIAA 2006-4911.

[42] P. March, "Woodward Effect Experimental Verifications," in AIP Conference Proceedings, 2004, Vol. 699, pp. 1138-1145.

[43] D. Bock, C. Drobny, P. Laufer, M. Kössling, and M. Tajmar, "Development and Testing of Electric Propulsion Systems at TU Dresden," in 52nd AIAA/SAE/ASEE Joint Propulsion Conference, 2016, AIAA 2016-4848.

[44] R. Shawyer, "Second generation EmDrive propulsion applied to SSTO launcher and interstellar probe" Acta Astronaut., Vol. 116, pp. 166-174, 2015.

[45] D. Brady, H. White, P. March, J. Lawrence, and F. Davies, "Anomalous Thrust Production from an RF Test Device Measured on a Low-Thrust Torsion Pendulum," in 50th AIAA/ASME/SAE/ASEE Joint Propulsion Conference, 2014, AIAA 2014-4029.

[46] H. White et al., "Measurement of Impulsive Thrust from a Closed Radio-Frequency Cavity in Vacuum," J. Propuls. Power, Article in Advance, Nov. 2016.

Discussion

During Martin's first talk, he mention's an electret capacitor, with asymmetric charge, as a possible way of getting a negative mass

Meholic: What would happen if you discharge an electret capacitor?

Tajmar: Well, I can't exactly discharge it, because to discharge I would need to connect the capacitor to a conductive circuit, and the electret is made of an isolator. So I have an isolator, and I'm bombarding it with ions or electrons which just stick to the surface. There is no current flow.

Meholic: How are you going to extract the usefulness of the negative mass out of that electret construct?

Tajmar: That's coming up in the next slides. By the way, if I get this to work I'll have a negative mass I can walk around with and I can sell it by the negative kilogram!

... audience laughter....

Fearn: During the Weber Electrodynamics section of the 1st talk, Martin describes the Wilhelm E. Weber force law, which just depends on charges, their separation, and velocity. It was a good enough description to derive the speed of light. It appears that Weber's force law does not take into account radiation reaction, which is very tiny and may have been overlooked at the time. Weber may not have known about it.

Tajmar: Yes, I'm coming to that, there was a later extension by A. K. T. Assis which adds in additional terms. Also, it turns out that massless charged particles don't radiate, this is apparently a new research topic.

Martin starts to talk about EM-drives...

March: You should treat this as an RF system, not an analog audio system. You need to have a dual directional coupler to your RF source and the test article. You need to look at the reflected power from the cavity and use the minimum of the SWR power tracker as a frequency tracker with an arbitrary ± offset.

Tajmar: That would be the ideal way to do it, and that's what we will try to implement next year. Certainly tracking the frequency is something that needs to be done and we have not set that up yet.

Martin starts to talk about Woodward's Mach Effect thruster work. Martin has an old thruster, Jim gave him from 1999, that he has started to run tests on. The new devices Jim runs only requires one frequency, the older devices needed two frequencies to be present, since they did not have electrostriction.

Rodal: The usual thing "now" is that Jim inputs an excitation frequency f, within f_{op}/Q_m bandwidth of the first natural frequency $f_{op} \sim 34$ KHz due to the piezoelectric effect, and that the electrostriction of the material naturally provides an excitation at $2f$, twice the excitation frequency f. However, note that the electrostiction resonance occurs at $(1/2)f_{op}$, at half the piezoelectric natural frequency f_{op} so that $2f = f_{op}$ and that the electrostriction resonant amplitude is orders of magnitude lower amplitude than the piezoelectric resonance.

Woodward: It's more complicated in this case José, because the thruster that Martin is checking is not like the ones that Heidi and I are running now, or like the ones tested by Nembo and George (they have newer devices). We are all using devices based on the Steiner–Martins SM-111 material, which has electrostriction as well as exhibiting the piezoelectric effect. The stack that Martin has, is made of EDO corporation (an American company now acquired by ITT corporation in 2007) EC-65 material discs. I don't know if that has any electrostriction response so he has to input two frequencies. It's a soft PZT material with a high dielectric constant of around 5000, it has about 4% dissipation. I built the stacks out of this stuff back then (1999) because it was cheap, they were a gift...

...audience laughter...

Woodward: Martin has shown his preliminary results, that show a small thrust from the old 1999 device. This was the first measurement of a self sustained system, with power and amplifier on board the torsion balance, to show thrust, with a very high resolution.

...audience applause...

March: Jim, didn't one of your early papers have a prediction for the thrust level in these older devices?

Woodward: No, not a paper that I recall, but there may be a graph in my "Making stargates and starships" book, that plots a thrust curve against various input power levels. Usually these devices had a small thrust measured in μN.

Tajmar: We were expecting μN or sub-μN levels of thrust, and that is what we saw in this preliminary data.

Woodward: Your data clearly shows the switching transients, tomorrow I'll show you what happens when you switch DC power on/off ... that is to say the switching transients go away. Thank you Martin !

Tajmar: You're welcome.

Martin is talking about his first data sets for the EM drive that his students built...

Rodal: Why does the thrust increase from 15 to 40 seconds?

Tajmar: Well I believe in this case, it is simply a shift in the center of gravity as the copper cavity expands. So it is an artifact of the thermal expansion of the copper. When I turn off the power, the force stops, you see the displacement sensor shifting down slowly, as the copper cools off. But this cannot be a force, since the power is off.

Williams: You said at the end that you could not confirm the existence of thrust for the EM drive, why is that?

Tajmar: When my "null" measurement, (which is in a direction perpendicular to the forward and backward direction) gives me the same thrust reading as a forward (or +) force direction measurement, then I know I have reached the level of resolution of my experiment. I cannot then say for sure that what I have seen is real or some noise. I need to improve my experimental setup (next year) and try again with higher resolution.

Broyles: Are you planning to change the design of your EM drive in the test run next year? If so, what design are you planning to use?

Tajmar: That's partly why I am here at this workshop. I wanted to ask if this or that is a good idea to try... we need to learn from each other, to avoid repeating the same mistakes.

Editor's Note: Due to a scheduling constraint, Dr. Brandenburg's session was the evening before the formal start of the meeting, before the recording equipment was in place. Since his session was not recorded, his paper follows without discussion.

The GEM Theory of Energy and Momentum Exchange with Spacetime, and Forces Observed in the Eagleworks Q-V Thruster

John E. Brandenburg
Morningstar Applied Physics, LLC

The basic premises and results of the GEMS (Gravity-EM Super) Unification theory are presented as well as its application to EM or "Q-V" thruster results. The GEMS theory began as an attempt to unify the long-range forces of nature, gravity and EM. But it unexpectedly also yields the observed masses, spins and charges of the πmesons, which carry the Strong Force; as well as the W and Z boson of the Weak Force, thus unifying the four forces of nature in one theory for the first time. A new spin-zero, neutral particle, is predicted by the GEMS theory, of rest-mass 22 MeV. The GEMS theory is based on two postulates:

1. that gravity fields exist as arrays of $E \times B$ drift cells, or Poynting vectors

2. that gravity and EM forces separated in a correlated way with the separation of protons and electrons from the Planck scale after the deployment of a Kaluza-Klein compact 5th dimension.

The theory, to first order, assumes transfer of particle momentum to a rigid spacetime structure from the $E \times B$ fields of the array. Likewise, in modeling the EM thruster, the reaction to the action of the thrust is considered, to first approximation, to be transferred to the nearby large masses via a rigid spacetime. A linear theory is found, approximately yielding the observed thrust to power relation observed experimentally in the Q-V thruster experiment. At higher powers, a nonlinear effect is seen theoretically, which yields the approximate thrust to power relation seen in higher power (kW) experiments of 0.2 N/kW. In the nonlinear limit, the thrust goes as the applied EM power squared, so large levels of thrust can be expected theoretically. A simple calculation shows that rapid trips to Mars can be effected by using this EM drive, powered by large, megawatt-scale solar panels.

1. Introduction

The Q-V Thruster [1] appears to create a force due to an interaction between applied RF power and the vacuum itself, within a specially shaped container. This result, confirming experimental results obtained elsewhere, may represent a breakthrough in space propulsion. A conceptual model has been proposed based on an interaction between the RF and virtual particles whose presence is required by quantum theory. This device was built to try to reproduce the results of experiments by Shawyer [2], where microwaves of much higher power were directed into a closed asymmetric vessel and generated thrust at a level of 0.1 N/kW. This result has been reproduced in the Q-V Thruster experiment, albeit at much lower power levels and much lower thrust per unit power (Figure 1). The thrust detected by the device is a reaction force to momentum that is transferred to the virtual particles. Two problems are present in the Q-V results: one is the global conservation of momentum, and the other is the problem of the divergence-free nature of the vacuum EM field that would seem to preclude transfer of momentum to the virtual particles.

However both of these problems can be solved by considering that the Q-V thruster and other similar devices are exchanging momentum directly between EM fields and space-time itself, which to first order acts like a rigid background. This effect occurs in the GEMS theory because, in that theory, the fabric of space-time itself is electromagnetic and EM fields can interfere constructively and destructively to change the structure of space-time. In this brief manuscript, the basic GEM theory will be briefly presented and its application to the Q-V thruster, to explain the origin of the measured forces with their approximate magnitude and scaling with applied power.

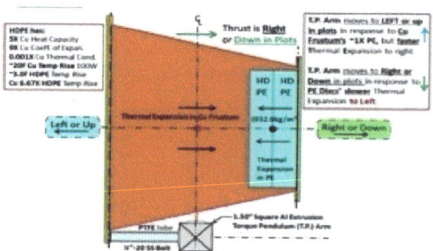

Figure 1: The Frustum of the Q-V Thruster

Based on the positive result of the Q-V thruster at low power, and the support for this result seen in the GEM theory, plus the great promise of this possible new means of space propulsion, a simple calculation will also be performed to see how long a 50-metric-ton, 1 megawatt solar-powered craft will take to go from Earth to Mars.

2. Momentum and Energy between Vacuum EM fields and Spacetime

In the standard theory of General Relativity the EM energy density

$$u = \frac{1}{2}\left(\epsilon_0 E^2 + \frac{B^2}{\mu_0}\right) \quad (1)$$

Expressing the EM energy density in terms of a mass density:

$$\rho = \frac{u}{c^2}$$
$$\nabla \cdot \mathbf{g} = -4\pi G \rho \quad (2)$$

In a plasma, the charged particles of the plasma would move to create currents to generate a $J \times B$ force to counteract this gravity force. But in a vacuum this is not possible. What then counteracts the gravity pull on the magnetic lines of force? We can answer this by going to the covariant form of the problem in general relativity. By starting with the problem of a EM field in a vacuum we can write

$$R_{\mu\nu} - \frac{1}{2} g_{\mu\nu} R = \frac{8\pi G}{c^4} T^{EM}_{\mu\nu} \quad (3)$$

where $T^{EM}_{\mu\nu}$ is the EM Stress tensor.

In covariant formalism we take the divergence of both sides and obtain, because the divergence the left side must vanish mathematically due to the Bianchi identities,

$$0 = \frac{8\pi G}{c^4} T^{EM\nu}_{\mu;\gamma} \quad (4)$$

Using covariant formalism, we have in expanded form

$$0 = T^{EM\nu}_{\mu,\gamma} + \Gamma^{\nu}_{\beta\gamma} T^{EM\beta}_{\mu} - \Gamma^{\beta}_{\mu\gamma} T^{EM\nu}_{\beta} \quad (5)$$

where $\Gamma^{\eta}_{\mu\nu}$ is the Christoffel symbol. The Christoffel symbol provides the part of the divergence that is due to gradients in the metric of space time, that is, gravity. Therefore, we obtain

$$T^{EM\nu}_{\mu,\gamma} = -\Gamma^{\nu}_{\beta\gamma} T^{EM\beta}_{\mu} + \Gamma^{\beta}_{\mu\gamma} T^{EM\nu}_{\beta} \quad (6)$$

This can be interpreted physically as spacetime, to first approximation, behaving as a rigid background. In the Newtonian limit for time constant gravity fields and using 3 vectors this becomes

$$-\frac{1}{c^2}\frac{\partial S}{\partial t} + \nabla \cdot T = \rho \quad (7)$$

where S is the Poynting vector. Careful summing of effects leads to the relation that $\rho = 2u/c^2$, or twice the expected mass density. It is for this reason that the angular deflection of starlight by the Sun is twice what would be expected from Newtonian theory.

Because we have a vacuum EM field with no charge,

$$\nabla \cdot T = 0 \tag{8}$$

Therefore we must have, to conserve momentum with $\rho = 2u/c^2$

$$-\frac{1}{2u}\frac{\partial S}{\partial t} = \mathbf{g} \tag{9}$$

This is the fundamental relation of the GEM theory, equating gravity to an $E \times B$ flow.

3. GEM theory and the Vacuum Bernoulli Effect

The Poynting vector is a fundamental quantity in EM theory and transports momentum and energy in EM fields. For example: a beam of light travels through space-time as a transverse electromagnetic wave expressed as the Poynting vector **S** as:

$$S = \frac{1}{\mu} E \times B = E \times H \tag{10}$$

This operation propels fundamental information about the elementary perturbation of space-time across the universe. The E and B fields expressed above are shown through the fundamental Poynting vector equation to be coupled at the point where the Poynting vector exists and couples to particles and space-time (Figure 2).

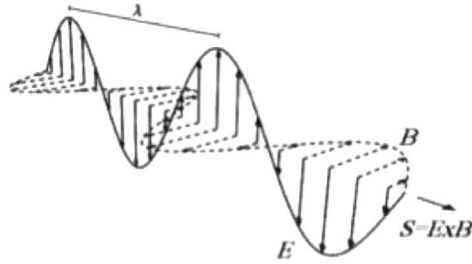

Figure 2: Components of a transverse light wave noting the propagation due to the Poynting vector

The Poynting fields around the "Morningstar Energy Box" [3] device can be visualized as seen in Figure 3 and are in the form of generating an electromagnetic vortex. Following our fluid concept of space-time, we can imagine that since the fluid space-time is stationary far away from the center of the Poynting vortex, a velocity gradient must exist. Such velocity gradients lead to turbulence when they exceed a small threshold, as is seen in everyday fluid flows. Added to this effect is the nonlocal nature of the wave functions of the particles, which sample the Poynting field at many locations at once, and thus do not see the vortex as a coherent entity but as a collection of interactions. So we can assume that the quantum mechanical matter waves will experience the Poynting vortex as a source of turbulence.

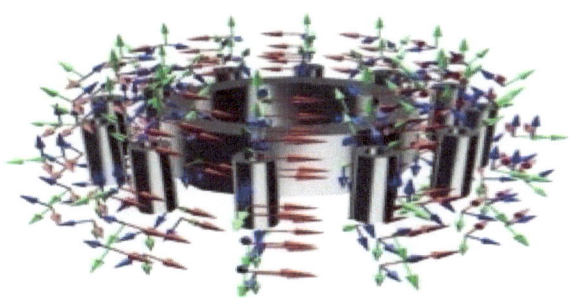

Figure 3: The electromagnetic fields surrounding a rotating "energy box" array of magnets. Magnetic fields are shown in blue, electric fields are shown in green, and the Poynting vector is shown in red. Note that the Poynting vectors form a vortex pattern

This intersection of fields is expressed in the Murad-Brandenburg equation, a Poynting conservation equation, which treats the Poynting vector field as a wave field, and away from its sources can be written:

$$\mu_0 \left[\frac{1}{c^2} \frac{\partial^2 \mathbf{S}}{\partial t^2} - \nabla^2 \mathbf{S} \right] = 0 \qquad (11)$$

When near field source terms are included we have:

$$\left[\frac{1}{c^2} \frac{\partial^2 \mathbf{S}}{\partial t^2} - \nabla^2 \mathbf{S} \right] = \nabla \cdot \left[\epsilon_0 EE + \frac{1}{\mu_0} BB \right] + \nabla \times \nabla \times \mathbf{S} \qquad (12)$$

where it can be seen the vorticity of the Poynting vector, $\nabla \times \mathbf{S}$, is prominent.

Away from sources, Poynting fields can be considered as a chaotic sum of waves, moving through each other. The Murad-Brandenburg Equation is a result of standard EM theory, but we can move beyond this theory to extend this with the GEM (Gravity Electro-Magnetic) theory.

The GEM theory [2] is a combination of the Sakharov theory of gravity as consisting of radiation pressure (Figure 4). That is, gravity fields are an array of $E \times B$ drifts arising from the quantum ZPF (Zero Point Fluctuation), and from the Kaluza-Klein theory of EM gravity unification through a hidden 5th dimension.

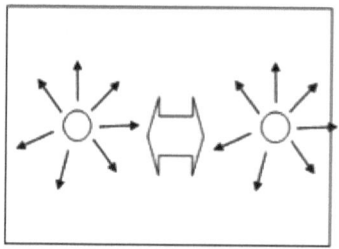
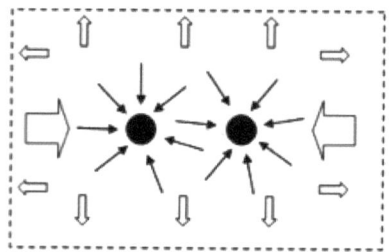

(a) Two bright objects in dark box repel each other

(b) Two dark objects in a bright box attract each other

Figure 4: The Sakharov model of gravity

It provides the basic mathematical results:

$$\ln\left(\frac{m_0}{m_p}\right) \equiv -(\alpha^{-1/2}+1)\ln\sigma \tag{13}$$

$$\ln\left(\frac{r_0}{r_p}\right) \equiv \sigma \tag{14}$$

where r_0 is the hidden dimension size, $m_0 = [m_p m_e]^{1/2}$, where m_p and m_e are the proton and electron masses respectively, $r_p = [G\hbar/c^3]^{1/2}$ is the Planck length, m_P is the Planck mass, and the square root of the mass ratio $(m_p/m_e)^{1/2} = \sigma = 42.8503$ is a parameter relating the electron and proton masses. This model has recently been refined to give corrected behavior near the Planck scale where all the quantities m_0/m_P, r_0/r_P, σ, and $\sigma \to 1$ leading to the corrected forms of Eq. 4a,b:

$$\ln\left(\frac{m_0}{m_p}\right) \equiv -(\alpha^{-1/2}+\alpha+1)\ln\sigma \tag{15}$$

$$\ln\left(\frac{r_0}{r_p}\right) \equiv \sigma - \frac{1}{\sigma^2} \tag{16}$$

The expression can be inverted to yield the formula for the Newton gravitation constant:

$$G = \left(\frac{e^2}{m_p m_e}\right) \alpha \exp\left(-2\left[\sigma - \frac{1}{\sigma^2}\right]\right) = 6.6752 \times 10^{-8} \text{ dyne } cm^2 g^{-2} \tag{17}$$

Which is within 2 parts per ten thousand of the measured value for G. And we find the proton mass, m_p, from the vacuum:

$$m_p = \sigma^{-(\alpha^{-1/2}+\alpha)} = 1.667 \times 10^{-24} \text{ g} \tag{18}$$

This result is within a 4 parts per thousand of the measured proton mass of 1.673×10^{-24} g. This demonstrates the importance and accuracy of the GEM theory in its developed form.

Recently, the GEM theory was able to predict the masses of the charged pion, W boson, and Higgs boson to high accuracy using the concept of quantum Mie scattering, or action integral, off of the structural resonances [2], of the classical EM radii, r_c. In a new analysis, which we briefly summarize here, this concept is generalized to include virtual paths of reduced probability of order α or $1/\sigma$:

$$\frac{E\ell}{c} = Nh \tag{19}$$

$$\ell = 2\pi r_c(1 + \alpha M) \tag{20}$$

where E is the particle rest energy, c, is the speed of light h is Planck?s constant, and ℓ is the path length. The previous derived masses were all for the case $N = 1$ $M = 0$. The most probable path is thus just simple circumference around a particle classical radii, but with a reduced probability, the path may divert to orbit the particle M times, this giving an effectively longer path. For the reduced probability cases of $M = 5$ and even $N = 5$, reflecting the dimensionality 5 of the GEM theory, we have the following particle masses, including a particle exclusively predicted by the GEM, the $M*$, (Morningstar) particle never before observed (see Table 1):

Table 1: **Particle masses predicted by the GEM theory and observed masses including the new predicted $M*$ particle**

Particle	Predicted Mass	Observed Mass	Error
Neutral Pion	135.12 MeV	134.98 MeV	0.1 %
Z Boson	91.03 GeV	91.19 GeV	0.2 %
Muon	105.63 MeV	105.66 MeV	0.02 %
M* Particle	21.9 MeV	****	****

Returning to the problem at hand, how to explain the Q-V thruster results, we begin by looking at the Poynting vector:

$$S = \frac{1}{\mu} E \times B = E \times H \tag{21}$$

$$S = \frac{1}{\mu_0} E \times B \tag{22}$$

Now the $E \times B$ drift will move all charged particles at the same speed and can be written in terms of S. For a vacuum we have, with u_0 as a steady state magnetic field energy density, the $E \times B$ drift velocity:

$$V = \frac{E \times B}{B^2} = \frac{S\mu_0}{B^2} = \frac{S}{2\mu_0} \tag{23}$$

This $E \times B$ velocity depends only on the ratio E/B and not on the mass of the particles affected or their charge. As a practical matter, the particles only assume this $E \times B$ motion after a cyclotron period, but we assume they are all "up to speed". We can adopt the physical model that E/B is the speed of the quantum vacuum since it obeys the equivalence principle and effects all masses the same. This means we can assume all the quantum particles appearing and disappearing from Heisenberg Uncertainty move at this rate. We can keep the magnetic field constant and create a gradient in the E field by tilting the plates relative to each other while keeping the E field everywhere normal to the B field as seen in Figure 5. This model has been tested and verified with a particle simulation code for the curvature E and B field configuration and the results are shown in Figure 6.

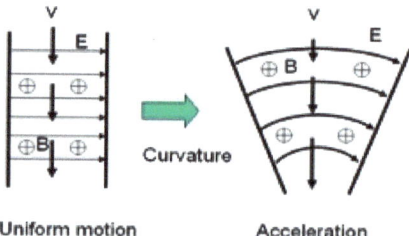

Figure 5: Motion of charged particles in cross E and B fields, with E vector formed between charged plates and B vector coming out of the paper. In the second case using tilted plates the charged particles accelerate. Velocity for all particles is the same regardless of charge or mass.

When this model of EM gravity is combined with Poynting's theorem, the Kaluza-Klein action falls out as a conserved quantity and can be called the Vacuum Bernoulli Equation (VBE) [4]. A brief version of it derivation shown below. We assume B^2 is constant and vary E in time, then the charged particles will all accelerate at the same rate:

$$\dot{V} = \frac{\dot{E} \times B}{B^2} = \frac{\dot{S}}{2u} \tag{24}$$

We can also write form Newtonian gravity theory with gravity vector field g, where G is Newton's gravity constant

$$\nabla \cdot \mathbf{g} = -4\pi G \rho \tag{25}$$

where we assume $E = mc^2$ and so an EM energy density can form a mass density as a source for a gravity field. This density ρ becomes:

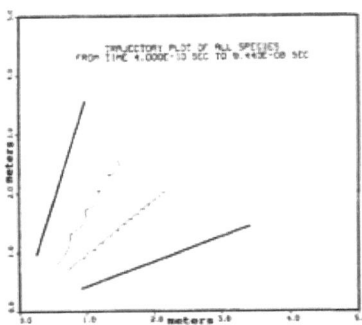

Figure 6: A Particle code simulation of the $E \times B$ drift gravity model showing an electron and a 10x electron mass positron

$$\begin{aligned} \rho &= \frac{u}{c^2} \\ u &= \frac{1}{2}\left(\epsilon_0 E^2 + \frac{B^2}{\mu_0}\right) \end{aligned} \quad (26)$$

This means when EM energy flows into a spherical region from all sides, gravity vectors pointing into the region increase in time so that, for the case of a spherically symmetric region, we have:

$$\nabla \cdot \dot{\mathbf{g}} = -4\pi G \dot{\rho} = -\frac{4\pi G}{c^2}\nabla \cdot \mathbf{S} \quad (27)$$

where both vectors can generate an additional vortex-like field $F = \nabla \times A$ that include curls of a vector potential.

For the simplest case of no "curl fields" we have,

$$\begin{aligned} \frac{\dot{\mathbf{g}}}{4\pi G} &= \frac{\mathbf{S}}{c^2} \\ \mathbf{g} \cdot \frac{\dot{\mathbf{g}}}{4\pi G} &= \mathbf{S} \cdot \frac{\dot{\mathbf{S}}}{2u_0 c^2} \\ \frac{g^2}{4\pi G} &= \frac{S^2}{2u_0 c^2} \\ \frac{g^2}{2\pi G} - \frac{S^2}{u_0 c^2} &= 0 \end{aligned} \quad (28)$$

This is the VBE expression we get from the Kaluza-Klein action in the Newtonian limit, with $\langle E \cdot B \rangle = 0$ in the vacuum, that is, a vacuum made of EM waves.

$$\text{KALUZA KLEIN ACTION} = \frac{R}{16\pi G} - \frac{F^{\mu\nu}F_{\mu\nu}}{4} \rightarrow \frac{g^2}{2\pi G} - \frac{S^2}{u_0 c^2} = 0 \quad (29)$$

Therefore, the same $E \times B$ drift theory of gravity, EM fields directly effecting spacetime rather than merely serving as a mass density source term, is also the basis for the coupled equations of General Relativity and Electromagnetism [5].

The Vacuum Bernoulli Equations says that gravity fields are associated with a net Poynting Flow in the vacuum. Therefore, we can change the local gravity field by changing the Poynting fields.

Now, we perturb the Poynting flow with a new an artificial Poynting flow, in the case of the Q-V thruster, created by the applied RF field. This perturbing flow is at right angles to the main Poynting flow and assumed of equal magnitude and is due to photon-photon scattering [6], a commonly observed phenomena, so the two flows can have a constructive interference term $d\mathbf{S} \cdot \mathbf{S}_\perp \approx |dS||S|$.

$$\frac{d\mathbf{g} \cdot \mathbf{g}}{2\pi G} = \frac{d\mathbf{S} \cdot \mathbf{S}}{uc^2}$$

$$\frac{|dg|}{|g|} \frac{g^2}{2\pi G} = \frac{d\mathbf{S} \cdot \mathbf{S}_\perp}{|S^2|} \frac{S^2}{u_0 c^2} = \frac{|dS|}{|S|} \frac{S^2}{u_0 c^2}$$

$$\frac{|dg|}{|g|} \cong \frac{|dS|}{|S|} \tag{30}$$

Now since we can assume each Poynting or $E \times B$ flow S is a "flow of the vacuum" and all it contains, and that it is a continuous flow field, we can perturb the flow fields as though they are of comparable underlying energy. We will assume the flow rate of the vacuum at the Earth's surface to be the escape velocity $V_{esc} = 1.1 \times 10^4$ m/sec, since that is the velocity of a particle falling from outer space. We will call this the assumption that "all vacuums are weightless", which is an extension of the equivalence principle to the vacuum itself, and says we can combine their $E \times B$ flows.

4. The Newtonian Gravity Potential

We have then a gravity potential in terms of an $E \times B$ drift model of gravity that is valid for both DC and oscillating E fields, where charged particles are accelerated into the strongest part of the perturbing E field. How then does the Newtonian gravity potential between charged particles come about? We begin with the expression for a gravity potential in terms of E and B fields in the vacuum, where V_D is the particle drift velocity in the crossed E and B fields. Here we use esu units for electromagnetic quantities:

$$\langle g_{00} \rangle = -1 - \frac{2\phi}{c^2} = \frac{E^2}{E^2 - B^2}$$

$$E^2 = E_0^2 \text{ or } E_1^2$$

$$-1 - \frac{2\phi}{c^2} = -1 - \frac{E_1^2}{B^2}$$
$$\frac{\partial V_D}{\partial t} = V_D \frac{\partial V_D}{\partial x} = \frac{Ec^2}{B^2} \frac{\partial E}{\partial x} \tag{31}$$

We now consider the mechanisms of how gravity arises from our $E \times B$ drift model and the interaction of charged particles with the quantum vacuum. We obtain the Newtonian potential as the perturbing E electric energy density divided by the powerful ZPF magnetic field:

$$\phi = \frac{1}{2} \frac{\langle E_1^2 \rangle}{B_0^2} c^2 \tag{32}$$

Note that this is expression for the gravity potential.

We can now proceed approximately with the derivation of the Newtonian potential from the GEM model of gravity potential shown in Figure 5 as an array of $E \times B$ drifts. We can consider the bending of light by gravity to be photon-photon scattering (Figure 7).

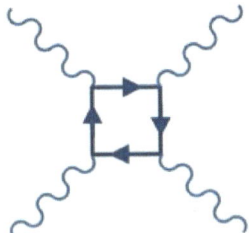

Figure 7: A Feynman diagram of photon-photon scattering, a well known process in the quantum vacuum.

According to the Standard Model all massive particles, electrons and quarks making up ordinary matter, are charged point particles. These charged particles all move freely in the presence of the ZPF fields of the quantum vacuum. It has been pointed out by Puthoff [7], that under the Standard Model even the quarks move freely because of the phenomenon of "Ultraviolet Freedom" and hence their interaction with the quantum vacuum can be considered in isolation. All these free charged particles are in constant motion, "Zitterbewegung", or quantum jitter, because of their accelerated motion must radiate as discussed by Puthoff. The radiation field is irregular, but statistically isotropic. The radiation E field is normal to the radiation direction coming from the particle and decays as $1/r$, where r is the distance from the particle. This radiation field constructively interferes with a portion of the ZPF that is isotropic and uniform, to surround the particle, resulting in an electric field energy density. It is this electric energy density that forms the numerator of the fraction. The magnetic energy density of the ZPF is the denominator of the fraction. Using our expressions for the classical radius of a charged particle, the Planck length and writing G as $G = c^4/(T_o r_p^2)$, we can write, using $B_o^2 = T_o$

$$\frac{1}{2}\frac{E_1^2}{B_0^2}c^2 = \frac{Gr_p^2}{2c^4}E_1^2 \tag{33}$$

The particle radiating because of its motion in the ZPF creates an electric field stress on the surface of a sphere of radius r, centered on the particle, and is proportional to the radiated power of the particle, where a is the acceleration of the particle.

$$4\pi r^2 \frac{E_1^2}{8\pi} c = \frac{2}{3}\frac{e^2}{c^3}a^2 \tag{34}$$

This expression is limited to $a < c^2/r_c$, where $r_c = e^2/mc^2$ is the particle classical radius in esu units. We limit the acceleration to the value $a = c^2/r_c$, and obtain, upon simplification:

$$E_r^2 = \frac{4}{3}E_c^2 \frac{r_c^2}{r^2} \tag{35}$$

Where $E_c = e/r_c^2$, the electric field at the classical particle surface. We then write the mean constructive interference term between the particle radiation and the background ZPF fields where $E_o = q_p/r_p^2$ takes into account the geometrical variations and time fluctuations, and obtain approximately:

$$\langle E_r \sin(2\pi\nu t)\rangle \sim \frac{1}{2\pi}E_r \tag{36}$$

where $\nu = 1$. From this expression we then obtain:

$$\langle E_r E_0 \rangle \cong \frac{1}{2\pi}\left(\frac{4}{3}\right)^{1/2} E_c \frac{r_c}{r}\frac{q_p}{r_p^2} \tag{37}$$

Gravity fields arise in the GEM theory from the constructive interference of the action of the ZPF:

$$\frac{1}{2}\frac{\langle E_r E_0\rangle}{B_0^2}c^2 \cong \frac{c^2}{2\pi}\left(\frac{4}{3}\right)^{1/2}\frac{e}{rr_c}\frac{q_p}{r_p^2}\frac{Gr_p^2}{c^4} \tag{38}$$

Using the expression for the Planck charge $q_p = e\alpha-1/2$, where α is the fine structure constant, we simplify Eq. (29) and obtain:

$$\frac{1}{2}\frac{\langle E_r E_0\rangle}{B_0^2}c^2 \cong \frac{\alpha^{-1/2}}{4\pi}\left(\frac{4}{3}\right)^{1/2}\frac{Gm}{r}$$

$$\frac{\alpha^{-1/2}}{4\pi}\left(\frac{4}{3}\right)^{1/2} = 1.07$$
$$\frac{1}{2}\frac{E_1^2}{B_0^2} \cong \frac{Gm}{r} \tag{39}$$

Thus, the Newtonian gravity potential can be recovered, to within factors close to one, from a physical model of $E \times B$ drifts of particles in a combination of the fluctuating fields of the particles radiation in response to the ZPF, and fibrous magnetic flux and fluctuating E fields of the ZPF. The presence of the charged particle breaks the symmetry of the spacetime and causes a $1/r$ electric field energy density to form. The gravity force is thus not a steady force on an individual particle but an average acceleration in this model. The weakness of gravity, caused by the smallness of G, is due to the strong nature of the ZPF magnetic fields. The $1/r$ dependence of the potential stems from the $1/r$ dependence of the radiation fields of the jittering particle, constructively interfering with the uniform background of the ZPF electric field fluctuations. These effects are, of course very small. However, the radiation field interference terms are independent for each particle and can add, causing the gravity force to combine in large ensembles of particles in a way that the pure EM force cannot. The gravity force can thus be said to be the result of the statistical mechanics of the fields of charged particles interacting with the vacuum around them, and combining in large ensembles.

Let us assume in the frustum that the EM waves follow the pattern of the simulations and create a concentration of field near the large end of the frustum. We will assume here, as in our derivation from the principle of a massless vacuum, that the magnetic field need not be that of the EM waves but is a magnetic field from the ZPF.

$$\phi = \frac{1}{2}\frac{\langle E_1^2 \rangle}{B_0^2} c^2 \tag{40}$$

Using the model of the gravity potential as created by a gradient of E^2 in a uniform background B field we find that the inclusion of plastic disks in the small end of the frustum suppresses the E field in that region. Thus the region near the wide end of the frustum has much more E field than the small end even without plastic dielectric disks, but that the inclusion of the disks in the small end will amplify the E^2 gradient. In the GEM theory this will create a curvature of space-time creating a gravity field pulling on the large end of the frustum and thus pulling the frustum towards the small end (Figure 8).

We can estimate the magnitude of the force via the GEM theory by using the vaccum Bernoulli equation.

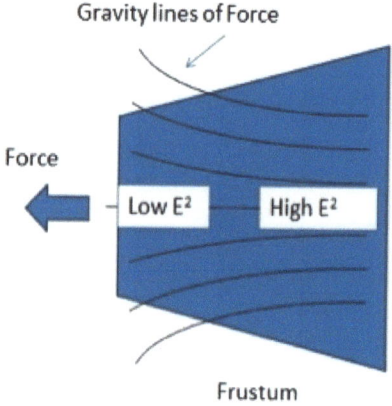

Figure 8: The gradient of E^2 caused by the standing EM fields in the frustum create, via the GEM theory, a curved metric and thus a net gravity force on the frustrum.

5. Action-Reaction and Momentum Conservation in the GEM theory of the Q-V Thruster

In a normal plasma thruster, real particles are accelerated by EM fields to depart the thruster and this gives a reaction force in agreement with Newton's 3rd law of motion. The reaction force accelerates the thruster, and the spacecraft it is attached to will give an equal and opposite momentum to the exhaust. If we did not use plasma but merely radiated microwaves out of an open metal vessel instead, this would also give a reaction force, albeit a small one per unit of power expended, because the EM waves carry momentum via the Poynting vector. However, by standard EM theory, if the metallic vessel is closed, the EM waves cannot escape and instead bounce around in the vessel, exchanging no net momentum with the walls, thus producing no net Poynting flow and thus no thrust. However, standard EM theory must be modified to include GEM effects, the fact that spacetime is electromagnetic, and thus can carry momentum itself. In the case of the frustum, the intense, and asymmetrically distributed, EM fields inside can modify spacetime, inducing a space-time curvature, and thus create gravity fields that create a net force on the Frustum.

Interpreting this thrust as a reaction force, where is the corresponding action? Stated differently: what then is this a reaction force to? The force on the frustum, that must exist to satisfy Newton's 3rd law? What if the device freely accelerated in space? Where would the momentum be that balanced its acceleration? The answer from the GEM theory is that the force on the frustrum occurs because of the GEM interaction with the gravity field (curved spacetime) of the Earth and thus the frustrum is pushing against the Earth via its gravity field. By this analysis, a spacecraft propelling itself by a Q-V thruster away from the Earth would cause the Earth to recoil. This is because gravity fields, even in the Newtonian limit, transfer momentum like EM fields.

A simple example of gravity fields exchanging momentum with EM fields is the bending of light by gravity fields (Figure 9). Obviously, the momentum carried by the light ray is changed, the global momentum flow is then to deposit the reaction to this exchange of momentum to the mass creating the spacetime curvature.

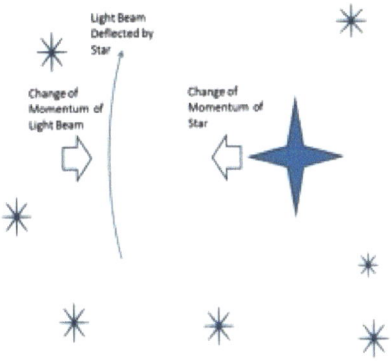

Figure 9: The exchange of momentum between a light beam and a nearby star. The star must provide "reaction" required by Newton's 3rd law in order for momentum to be conserved.

6. The Thrust Versus Power Relation

To complete this calculation we need to estimate S in the gravity field at the Earth's surface. The GEM theory says that gravity is essentially an EM interaction at the subatomic scale, and so we can write the gravity force acting on each nucleon as a radiation pressure acting on an EM cross section that is proportional to mass. The GEM theory allows us to write for a nucleon in the Earth's gravity field:

$$P_{\text{EM}}\sigma_n \cong m_n g \tag{41}$$

where $\sigma_n \approx 10^{-26} cm^2$ is the EM cross section of a nucleon, similar to the Thompson cross section of an electron, and m_n is a typical mass of 1 amu $= 1.7 \times 10^{-24}$ g. This model is aided by the fact that nuclear matter occupies a fixed volume per unit mass, so individual nucleons preserve their size in a nucleus:

$$P_{\text{EM}} \cong \frac{m_n}{\sigma_n} g \qquad (42)$$

The mass per unit area is then $m_n/\sigma_n \approx 50\ g/cm^2$ or $500\ kg/m^2$, surprisingly similar to macroscopic matter. Using g at the Earth's surface we obtain $P_{\text{EM}} \approx 5 \times 10^3\ J/m^3$.

Outer space vacuum is thus arriving at $V_{\text{esc}} = 1.1 \times 10^4\text{m/sec}$ and we can write:

$$S = V_{\text{esc}} V_{\text{esc}} \cong 6 \times 10^7\ W/m^2 \qquad (43)$$

The perturbing Poynting flux, which we assume is asymmetrically absorbed in the wall nearest the field concentration, on the large end of the frustum (area approximately 0.1 m^2), is approximately $P/A = 500\ W/m^2$. Thus, we can write for the steady-state perturbation of space time curvature due to the asymmetric S field in the thruster:

$$F_{\text{QV}} \cong m_{\text{QV}} g \frac{dS}{S} \cong 1 \times 10^{-5} N = 10 \mu N \qquad (44)$$

We can also write this force as a function of applied RF power.

$$\frac{F_{\text{QV}}}{W} \cong \frac{m_{\text{QV}} g}{AS} \cong 2 \times 10^{-7} N = 0.2 \mu N/W \qquad (45)$$

in approximate agreement with the experimental results of $F_{\text{QV}}/W = 0.7\ \mu N/W$.

Therefore, the results of the Q-V thruster experiments and other similar experiments can be explained through the GEM theory. This GEM model is somewhat primitive, but can be refined with the help of more experimental data. This effect is inherently non-linear due to the presence of S^2 terms in the GEM equations, so the low thrust per unit power can be expected to improve at higher power densities such as employed in the Shawyer experiment.

The GEM interpretation of the Q-V data appears much different than the quantum virtual plasma model of the Q-V thruster but is actually very similar. Both models assume a reaction mass tied to the vacuum itself. In the case of the GEM theory, that vacuum is spacetime itself and is tied to the Earth and other nearby masses. In the case of the Q-V theory, it is the virtual particles that are part of the quantum vacuum, and must close the momentum transfer equation by transferring momentum through spacetime to nearby masses.

In the low power experiments Vacuum Bernoulli Equation is in effect in a linear perturbation model, being proportional to the applied power. However, at high powers we can expect the Vacuum Bernoulli Equation to enter into a fully nonlinear mode and the gravity force will be proportional the square of the power. This can be seen from the VBE with the assumption that he cavity will act like a high Q resonator, with high circulating power. Assuming an power of 1 kW and a Q =10,000 (typical for a copper resonator), we can assume a power flux of 10^8 W/m^2. In this case we have the equation

$$\frac{|dg|}{|g|}\frac{g^2}{2\pi G} = \frac{S'^2}{|S^2|}\frac{S^2}{u_0 c^2}$$

$$\frac{|dg|}{|g|} \cong \frac{S'^2}{|S^2|} \quad (46)$$

where S' is the applied circulating power of $10^8 W/m^2$ and S is the Poynting flux due to the Earth's gravitational field. We obtain then at 1 kW input power, the approximate thrust force,

$$F_{QV} \cong m_{QV} g \frac{S'^2}{|S^2|} \cong 0.1 N \left[\frac{100 MW}{60 MW}\right]^2 = 0.27\ N\ . \quad (47)$$

with a thrust that should increase as the power squared.

7. Application to Spaceflight: An Approximate Mars Mission Calculation

Using the 0.1 N/kW value from the high power Shawyer experiment [2], we can find a simple estimate for the total ΔV and trip time to Mars. Assume a 30 metric ton solar powered space craft with a power of 1 MW from a high performance solar array, which we will assume reconfigures itself to maintain constant power on the way to Mars. This gives a thrust of 100 N and an acceleration $T/M = 3.33 \times 10^{-3}$. Here we take advantage of the fact that a spiral out to Mars orbit at $R_M = 1.5$ A.U. from $R_E = 1.0$ A.U. (see Figure 10) involves a $\Delta R/R_E < 1$, and this trip can be expected to take place in much less than an Earth orbit period : $\Delta t/P_{orbit} \ll 1$. Thus, we can use the approximation that the trajectory spirals out through a series of orbits with the circular orbit condition,

$$\frac{GM_s}{R} = V_\theta^2 \quad (48)$$

where V_θ is the rotational velocity in the, R is the radius from the Sun, and M_S is the mass of the Sun. The total change in specific energy is approximately:

$$\Delta W = -\frac{GM_s}{2R_E} + \frac{GM_s}{2R_M} = 150\ km^2/sec^2 \quad (49)$$

However, because the the orbit will actually be a spiral outward and not a series of circles, part of the thrust will ineffective due to the thrust vector not being aligned with the rotational component of velocity. We can approximate this inefficiency by expression,

$$\Delta W \cong V_E \frac{T \Delta t}{M} < \langle \cos \phi \rangle . \qquad (50)$$

The average projection of the thrust vector onto the rotational velocity on the spiral orbit is a function of $\Delta R/\ell$, where we have defined the parameter $\ell = 2\pi R_E (\Delta t / P_{\text{orbit}})$, where P_{orbit} is the period of the original orbit. We obtain in the limit of $\Delta R/\ell$ and $\Delta t/P_{\text{orbit}}$ both $\ll 1$:

$$\langle \cos \phi \rangle \cong \frac{\ell}{\sqrt{(\Delta R)^2 + \ell^2}} \qquad (51)$$

This system gives a correct limit of $\langle \cos \phi \rangle = 1$ or zero gravity losses, in the limit of $\Delta R/\ell \ll 1$, a spiral out over many orbital periods for $\Delta R/R_E \ll 1$. Solving the system of Eqs. 46 and 47 by iterations, we obtain the estimate $\langle \cos \phi \rangle \approx 0.7$, for an average angle of the spiral of $\phi \cong 45°$. We then obtain by this analysis a $\Delta V \approx 7.2$ km/sec. This is roughly double the required $\Delta V \approx 3.5$ km/sec for a minimum energy Hohmann Transfer requiring a $\Delta t \approx 10$ months. This increase in ΔV for low thrust trajectories is due to gravity losses and is unavoidable [8]. However, despite the gravity loss inefficiency, the required for our Q-V thruster is $\Delta t \approx 4$ weeks or ~ 1 month, so we can take advantage of abundant solar power and the propellant-less character of the Q-V thruster to get to Mars in $1/10$ the time required for more conventional chemical fuel approaches. Accordingly, assuming the Q-V thruster results at high power can be reproduced, this propulsion technology will be a true breakthrough in space propulsion.

8. Conclusions

Creating thrust by injecting microwaves into an isolated asymmetrical metal container may seem impossible at first glance, but if one accepts the concept that spacetime is fundamentally electromagnetic, then forces on the asymmetrical container are not only possible but expected. Creating an asymmetrical EM field, under the GEM theory, will directly create a curvature in spacetime and thus a gravity force. The gravity force, a curvature in local spacetime whose structure connects all large masses in the vicinity, creates a force on the metal container, and it also creates a reaction force that conserves momentum with the Earth and Sun, that anchor the local structure of spacetime.

The GEM theory predicts that gravity fields are a distortion of the quantum ZPF fields and have a net Poynting flow. That is, the fabric of spacetime is electrodynamic, consisting of ZPF fields. This theory also

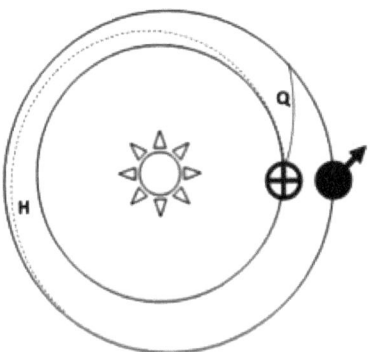

Figure 10: The approximately 1 month trajectory of a Q-V Thruster propelled spacecraft (path Q) versus a 10 month minimum energy trajectory Hohmann transfer trajectory for a spacecraft (path H) journeying from Earth to Mars.

predicts that we can change the Poynting flow associated with gravity by constructive and destructive inference between the ZPF and artificially applied Poynting flows. Thus, in the GEM theory Poynting flows can create artificial curvatures in the ZPF and by this curve spacetime creating local gravity fields that can create forces on a spacecraft or its components. The reaction force to this created force is felt by spacetime itself and transferred to the nearby astronomical masses such as the Earth. The medium for the transfer of momentum is the gravity field itself, as if it was a solid object. This is similar to the bending of light, an EM field, in a gravity field where EM momentum is exchanged with the gravity field. (Figure 10) Therefore, the application of EM Poynting field in carefully controlled geometries can, in the GEM theory, create gravity forces. This preliminary analysis suggests the frustum experiment at Eagleworks may be creating forces by bending spacetime, and the GEM theory allows the calculation of magnitude and scaling behavior to be made, and gives approximate agreement with what is observed.

In a very recent development, the GEM theory prediction of a neutral, spin 0, particle of mass-energy $M* = 21.9 \text{MeV}$ [9,10] that would decay into electron-positron pairs, has been partially confirmed with the discovery of a similar particle "X" at $M_x = 16.9$ MeV [11]. This newly discovered particle with a spin 0 and decaying into electron-positron pairs appears to be a perturbed state of the $M*$ particle, caused by the neighboring electron quantum mass state:

$$M_x \cong \frac{M*}{1+\alpha\sigma} = \frac{21.9\, MeV}{1.31} = 16.7\, MeV \qquad (52)$$

In summary, it appears possible that a great breakthrough in space propulsion has been made at a NASA associated laboratory. This experimental breakthrough may be explained, both conceptually and quantitatively, by a theory of Gravity-EM unification that has been in development for decades, and which yields the value of G and the proton mass from quantum vacuum quantities to high accuracy [9,10]. Therefore, the road appears open to great advances and revolutionary changes in human spaceflight, to include the entire solar system. Let us proceed with all diligence and discuss and explore these possibilities further.

References

[1] Brady, D., White, H. G., March, P., Lawrence, J., and Davies, F. J., ?Anomalous Thrust Production from an RF Test Device Measured on a Low-Thrust Torsion Pendulum?, presented at the 50th AIAA Propulsion Conf., July 2015.

[2] Shawyer R. (2008) "Microwave Propulsion-Progress in the EMDrive Programe", 2008 International Astronautics Conference IAC-08-C4.4.7

[3] J.E. Brandenburg (2012) "An Extension of the GEM Unification Theory to Include Strong and Weak Nuclear Forces and an Estimate of the Higgs Boson Mass", STAIF II Conference Albuquerque NM March 2012

[4] Murad, P. A., Boardman, M. J. and Brandenburg, J. E., "The Morningstar Energy Box- Part Redux", Journal of Space Exploration - STAIF II, 2012.

[5] Brandenburg, J.E. and Kline, J. K., "Application of the GEM theory of Gravity-Electro-Magnetism Unification to the problem of Controlled Gravity: Theory and Experiment". AIAA 98-3137, 1998.

[6] Witten, E., "Anomalous Cross-Section for Photon–Photon Scattering in Gauge Theories", Nucl. Phys. B120, (1977) 189.

[7] Puthoff, H.E., "Gravity as a zero-point fluctuation force", Phys. Rev. A 39, 2333, (1989).

[8] Goebel, M., Dan and Ira Katz. *Fundamentals of Electric Propulsion*. New York: Wiley, 2008.

[9] Brandenburg J.E. "The GEM Unification Theory ; Extending the Standard Model to Include Gravitation", Lambert Academic Publishing. June 2016.

[10] Brandenburg , J.E. , "The GEM (Gravity-EM)Theory : the Unification of the Strong, EM , Weak ,and Gravity Forces of Nature",

[11] Feng, J. L. et. al., Particle Physics Models for the 17 MeV Anomaly in Beryllium Nuclear Decays, http://arxiv.org/abs/1608.03591

Theory of the EM Drive in TM Mode Based on Mach-Lorentz Theory

Jean–Philippe Montillet
Ecole Polytechnique Fédérale
Lausanne, Switzerland

Various theories have recently emerged to explain the anomalous thrust generated by the controversial EM Drive [1,2]. This work proposes a model based on the theory of the Mach-Lorentz thruster [3]. The thrust is generated by the combination between the Lorentz force and the Woodward effect [4]. The development has been facilitated due the discussions with Dr. Jose Rodal and Prof. Heidi Fearn. In addition, our approach is only based on the results from the experiments in TM mode released by the NASA Eagleworks group [5,6]. The purpose of this communication is to improve our model using feedback from scientists and to some extends with the EM Drive community in order to point out weaknesses on some of our assumptions and to plan future campaigns of experimental tests.

1. Overview

Since the first experiment at the beginning of this new millennium, the EM Drive has been the focus of many critics from scientists and engineers. In addition, public debates have also contributed in casting doubts on this possible technology. However, the latest tests and measurements by various academics [7] and government agencies [5], which should have dismissed this technology once and for all, have confirmed the anomalous thrust generated by this device. This latest development has sparked new interests for this device, which could play a critical role in space exploration of our solar system [8]. Nevertheless, the ultimate goal remains the creation of a model of the EM Drive supporting the experiments.

In the last two decades, various theories have emerged to understand the thrust generated by the EM Drive. The author in [1] or [2] developed a theory based on the difference of radiation pressure forces on the end plates of the cavity. More recently, an explanation of the anomalous thrust has been supported by the introduction of the Unruh radiation [9]. Another theory [10] attempts to model this exotic propulsion engine based on the emission of paired photons expulsed through the cavity end walls and generating the recored thrust. In [11], the thrust is the result of a man-made gravitational field gradient taking place inside the cavity. Other emerging theories can be found online. Among all those theories, we are here only interested in the application of the theory of the Mach-

Lorentz thruster (MLT) [3] to the EM Drive. Note that the MLT is also called Mach Effect Gravitational Assist-drive (MEGA-drive). This theory is based on the Lorentz force coupled to the Woodward effect [12] in order to explain the anomalous thrust. The Woodward effect is relies on the Mach's principle, which defines inertia within general relativity theory [13], and demonstrates that inertia is caused by the gravitational interaction between an object and massive bodies in the distant universe. The Woodward effect describes a way to extract a linear force from an accelerating object which is undergoing internal deformation and mass-energy fluctuations. Momentum is conserved via the gravitational field. Experiments with capacitors and piezoelectric materials have reproduced the Woodward effect in laboratory environment [4].

Our model assumes that each element constituting the EM cavity (frustum), namely the two end plates and the conical wall, responds independently to the EM waves propagating inside the cavity and reflected on the walls. Each element is modelled with a capacitor in series with a resistance and in parallel with an inductor. The capacitor models the EM excitation phenomenon from the waves reflecting on the end plates in TM modes. Thus, the assumptions are from the EM excitation: 1/ creating surface currents on the surface of the walls; 2/ generating an EM energy density "stored" in the skin layer of the copper end plates (e.g., evanescent waves [14]). The capacitor charges and discharges instantaneously due to the creation and dissipation of the charges. If the capacitor is related to the EM excitation mostly due to the electric field, the inductor is then modelling the EM excitation with the magnetic field via the Eddy (Foucault) currents phenomenon [15]. The Eddy currents are loops of electrical currents induced within conductors by a changing magnetic field in the conductor, hence generated when the vector field and the cavity walls are intersecting. While the capacitor and inductor model two different phenomena, the additional strong assumption is that the capacitor should be the dominant effect when the electric field is perpendicular to the wall. However, if the electric field is parallel to the wall or if something prevents the EM excitation on the wall, the inductor should then be the dominant model. For example, when inserting some dielectric (e.g., High-density polyethylene (HDPE)) to one end (e.g., small end plate), it could prevent (partially) the creation of electric charges on this particular wall. The electric field is more attenuated than the magnetic field when passing through the dielectric field (i.e. electrical insulator properties [16]). Thus, we model this phenomenon by increasing the resistance in series with the capacitor.

Now, the current propagating at the interior surface of the conical wall (between the two end plates) is also going through the magnetic field generated inside the cavity, hence resulting in a Lorentz force. This force is the result of the integration on the whole interior surface. However, this force alone cannot be responsible for any movement of the cavity due to the conservation of momentum as explained further in this document.

Secondly, the MLT model is based on the assumption that the Woodward effect is generating the thrust and it is triggered by the Lorentz force. The variation of mass described in [4] is driven by the variation of EM energy density in the skin layer of the copper wall. Thus, the assumption is that the Woodward effect mostly relies on the capacitor model of the cavity wall.

The next sections describe the various steps of this model based on the TM010 experiments [5]. In order to facilitate the understanding of the overall model, an analogy between electrical circuits and Newtonian mechanics is made. We must state clearly that there are two different mechanisms which can be modelled with an RLC circuit in this work. The first mechanism is the response to the EM excitation of each element composing the cavity which basically explains two phenomena described above: Eddy currents from the magnetic field, and the surface current from the electric field. The second analogy with the RLC circuit is used to explain the anomalous thrust by modelling the whole cavity. This model is fully developed in the following sections. However, our analogy does not relate to the well-known model of a specific EM cavity with an RLC circuit used in the analysis of the EM properties. Readers interested in this analogy can refer to [17].

2. Some Equations and Discussions

2.1 Modelisation of the Three Steps: Electro-mechanics and Gravitational Coupling (EMG)

RC circuit

Let us first assume that there is no force or no thrust acting on the cavity. The electric field is exciting the end plates, and parallel to the conical wall (e.g., [18] or [19]). The capacitor models the EM excitation via the electric field on the end plates. Thus, the capacitor charges and discharges instantaneously due to the creation and dissipation of the charges by EM excitation on the surface of the end plates. The EM cavity can then be modelled as two capacitors in series charging/discharging instantaneously. Taking into account the dissipation intrinsic to the conductor properties, the cavity can be modelled such as a RC circuit. The equations read:

$$Ri + \frac{q}{C} = 0$$
$$R\partial_t q + \frac{q}{C} = 0$$
$$q(t) \sim q_0 \exp\left(-\frac{t}{RC}\right) \quad (1)$$

q_0 is the charge at $t = 0$. The equation of the charge $q(t)$ shows that the dissipation of the initial charge q_0 during the discharge time $\tau =$

RC. That is why we can understand it such as a *switch on- switch off* of the capacitor. To evaluate the discharge time τ, one can write the conservation of charge equation at the surface of the plates.

Let us consider the density of the charge $\rho(t)$, the conductivity of the copper σ and its permittivity ϵ_r, then [14],

$$\partial_t \rho(t) + div \vec{j} = 0$$
$$\partial_t \epsilon_r \epsilon_0 \vec{E} + div \sigma \vec{E} = 0$$
$$\epsilon_r \epsilon_0 \partial_t \Delta V + \sigma \Delta V = 0$$
$$V(t) \sim V_0 \exp\left(-\frac{\sigma}{\epsilon_r \epsilon_0} t\right)$$

(2)

The discharge time τ equal $\frac{\epsilon_r \epsilon_0}{\sigma}$ or $6*8.85e-12/5.85e7 \sim 1e-18s$ (values from [20]). Note that we assume at the surface of the plate $\vec{E} = -\vec{\nabla} V$ (no magnetic potential). V_0 is the potential at $t = 0$ before the discharge. Now in order to evaluate the potential over the whole copper end plate, we integrate on the whole surface S. The difference of potential between the two end plates (without dielectric or HDPE insert) is then $DV = (S_1 - S_2) V_0 \exp\left(-\frac{\sigma}{\epsilon_r \epsilon_0} t\right)$. Within the frustum model, S is equal to πr^2 (r the radius of the end plate). With the insertion of the HDPE (or dielectric) on the end with surface S_2, the difference of potential is then equal to $DV \sim S_1 V_0 \exp\left(-\frac{\sigma}{\epsilon_r \epsilon_0} t\right)$. Now, the Eddy currents generated on the conical wall due to the perpendicular magnetic field, compete with the current propagating from the difference of electric potential between the end plates (from large to small end plate). The direction of the Eddy currents depends on $\vec{curl} \vec{B}$ (see the Maxwell equation $\vec{curl} \vec{B} = \mu_0 \vec{j}$, with μ_0 the permeability of the vacuum and $\vec{j}$ the Eddy currents). The two currents propagate in opposite directions in the TM010 scenario. In addition, the Eddy currents may have a larger amplitude than the other current propagating on the conical wall.

In the first step, **the main assumption is the creation of charges at the surface of the end plates** in TM mode.

The acceleration of the cavity due to the Lorentz force

The second step is when a force is generated acting on the cavity. The current propagates inside the magnetic field, and thus triggering a Lorentz force F_{Lo}. As previously said, this current can be either the Eddy current or the current induced by the difference of electrical potential between the two end plates. Let us assume that an alternative current (AC) is propagating between the two end plates. In terms of circuit analogy, the cavity is now a RLC circuit with an induced electromotive force ε:

$$Ri + \frac{q}{C} + L\partial_t i - \varepsilon = 0 \tag{3}$$

$L\partial_t i$ is equivalent to the mechanical action of the cavity getting accelerated (or $m\partial_t v$ in classical mechanics (Newton's second law), m the mass of the cavity and v the speed). ε can be expressed such as $\varepsilon = -\partial_t \phi_B(t)$, with $\phi_B(t)$ the magnetic flux through the copper conical wall surface [14]. In classical mechanics (i.e. Newton second law), when projecting the forces on the Z-axis (see Figure 1), the equation (3) becomes:

$$m\partial_t^2 z = \alpha \partial_t z - Kz + F_{Lo} \quad (4)$$

where $\alpha \partial_t z$ is the dissipative force due to the resistivity of the copper when the current propagates. Note that the force due to the weight of the cavity is perpendicular to the axis onto we project the forces and the Z-axis direction is toward the small end plate.

Let us estimate the Lorentz force applied to one electron (with charge q_e and speed v_e) moving through the magnetic field $\vec{B}$ at the surface of the conical wall

$$\vec{F}_{Lo} = q_e \vec{v}_e \times \vec{B}$$

$\times$ is the vectorial product. Using the convention in [18] and [19], the magnetic field is parallel to the conical wall with only a component on the surface of the azimuth direction $\vec{B} = B_\phi u_\phi$. The apex angle of the frustum is defined as $2\theta_w$. The expression of the force on the Z-axis is then $\vec{F}_{Lo}$:

$$\vec{F}_{Lo} = q_e v_e B_\phi \sin(\theta_w) u_z$$

The displacement of the electrons is collinear to the unit length $\vec{dl}$ of the conical wall. If we assume that the number density of electrons in copper is n_{Cu}, dS the unit surface, then we can estimate the force over dl

$$\vec{F}_{Lo} = n_{Cu} v_e dS dl B_\phi \sin(\theta_w) u_z$$

Let us assume the current with an amplitude $dI_0 = n_{Cu} v_e dS$. Then the Lorentz force per unit of length dl is:

$$\vec{F}_{Lo} = dI_0 dl B_\phi \sin(\theta_w) u_z \quad (5)$$

Using the axis as defined in Figure (and the same as in [19]), (5) becomes:

$$\vec{F}_{Lo} = dI_0 B_0 exp(j(\omega t - Kz)) \cos(\theta_w) \sin(\theta_w) dz u_z \quad (6)$$

Note that the amplitude of the magnetic field at the surface of the conical wall is not constant and depends on the TM mode. In TM010, the experiments carried out by the NASA Eagleworks group, B_0 is constant in azimuthal plane [5], but not on the Z-axis. Thus, the Lorentz force can vary while the current propagates from one end to the other.

dI_0 can be integrated over the whole azimuth plane, but there is an assumption to be made: do we consider the current propagating over the whole thickness of the copper sheet, or just over an elementary part of it? It is important to underline that we are here using a simpel model of the Lorentz force applied to free charges in a conductor. However, because surface charges are distributed over some infinitesimal depth, and those charges at greater depths are shielded by the others and therefore see a smaller electric field $\vec{E}$. In other words, the electric field created by the displacment of those charges decreases in amplitude with the depth in the conductor. Moreover, we did not take into account the possible effect of Kelvin polarization forces [21]. Note that (4) is only stated for a pedagogical point of view, because a creation of thrust from this equation is prevented by the momentum conservation principle (i.e. special relativity).

In the second step, the main assumption is the **current propagating at the surface of the conical wall** inside the cavity, hence generating the Lorentz Force.

Generating the thrust

The last step is the triggering of the Woodward effect generating the thrust. Basically, it is the introduction of $\partial_t z \partial_t m$ into equation (4). As previously mentioned, the variation of mass of the cavity is due to the Woodward effect applied to the EM energy density *stored* in the skin layer of the copper end plate(s). Thus, the Woodward effect is mostly associated with the capacitor model and not the inductor for each element of the cavity, hence introducing a dielectric should reduce it. In TM010 mode, the effect should take place mostly on the end plates. Recalling the Woodward effect takes place only if the cavity is accelerated while the energy inside the cavity is fluctuating [4]. The variation of mass is translated into the equation [4],

$$\delta\rho_0(t) = \frac{1}{4\pi G}[\frac{1}{\rho_0 c^2}\partial_t^2 U_0 - (\frac{1}{\rho_0 c^2})^2 (\partial_t U_0)^2] \qquad (7)$$

U_0 is the energy of the system, ρ_0 is the transient mass source, and c speed of light. Considering a rest energy $\mathcal{E}$, energy of the frustum at rest, including all the particles within the frustum with no EM excitation, one can state the famous Einstein's relationship in special relativity between $\mathcal{E}$ and the rest mass ρ, $E = \rho c^2$. In Appendix III, we justify the assumption that the variation with time of $\mathcal{E}$ equal the variation of EM energy density with the capacitor model. The variation of EM energy in the copper end plate (skin layer) is expressed with du (see Appendix I,(??)). The Woodward effect in (7) can then be rewritten

$$\delta\rho(t) = \frac{1}{4\pi G}[\frac{1}{\rho c^2}\partial_t^2 u - (\frac{1}{\rho c^2})^2 (\partial_t u)^2] \qquad (8)$$

The author in [4] calls $\partial_t^2 U_0$ the impulse engine, and $(\partial_t U_0)^2$ the wormhole. In the next section, we discuss the quantities $\partial_t^2 u$ and $\partial_t u$ and possible explanations in terms of EM theory. Note that the reader can find the rigorous derivation of (8) (based on [4]) with the assumptions of replacing the input power with the electromagnetic energy density in the appendices.

Finally, we assume that the Woodward effect creates a variation of mass (mass density) independently for each end plate when considering $\mathcal{E}$ as the rest energy for one end plate in order to obtain (8). Let us then define:

- $\partial_t \rho L$: variation of mass at large end plate

- $\partial_t \rho S$: variation of mass at small end plate

with $\partial_t m = \partial_t \rho L - \partial_t \rho S$. (4) becomes

$$m \partial_t^2 z + \partial_t z (\partial_t \rho L - \partial_t \rho S) = \alpha \partial_t z - Kz + F_{Lo} \qquad (9)$$

One needs to underline that the terms $\alpha \partial_t z$ and Kz are intrinsic to the cavity parameters (i.e. resistivity, dimension), whereas the thrust or acceleration of the cavity ($m \partial_t^2 z$) depends on the Lorentz force F_{Lo} and the relativistic terms coming from the Woodward effect $\partial_t z (\partial_t \rho L - \partial_t \rho S)$. One can underline that $\partial_t \rho L - \partial_t \rho S$ can be interpreted as the Woodward effect created independently on each end plate with opposite direction (towards the outside of the cavity). Finally, the measurable thrust in the MLT comes from (9) which results from the coupling between the Lorentz force and the Woodward effect. Note that (9) sums up our model of the MLT.

In the last step, we assume that the **Lorentz force triggers the Woodward effect in order to generate the anomalous thrust.**

2.2 Variation of electromagnetic energy density

This section looks at numerical estimation of the EM energy density in the skin layer of the copper end plates.

Evanescent Waves in Copper Walls and Numerical Estimation

As seen in the previous section, the surface surcharges disappeared as soon as they are created (with $\vec{j} = \sigma_{Cu} \vec{E}$ and charge conservation equation, we have $\tau_{relax} = \frac{\epsilon_0}{\sigma_{Cu}} \sim 10^{-18}\ s \sim 0$). Note that in the following $\epsilon = \epsilon_r \epsilon_0$ and $\mu = \mu_r \mu_0$ as previously defined. We can then state the Maxwell equations at the surface of the copper wall end plates,

$$\left. \begin{array}{l} div \vec{E}_{tot} \sim 0, \\ \vec{curl} \vec{E}_{tot} = -\partial_t \vec{B}_{tot}, \\ div \vec{B}_{tot} = 0, \\ \vec{curl} \vec{B}_{tot} = \mu \epsilon \partial_t \vec{E}_{tot} + \mu \sigma_{Cu} \vec{E}_{tot}, \end{array} \right\}$$

The wave equation is then [14]:

$$\Delta \vec{E}_{tot} = \mu\epsilon \partial_t^2 \vec{E}_{tot} + \mu\sigma_{Cu}\partial_t \vec{E}_{tot} \qquad (10)$$

Assuming that the solution is a planar wave of the type $\vec{E} = \vec{E}_0 e^{i(\omega t - \vec{k}\cdot\vec{r})}$ ($i = \sqrt{-1}$), and knowing that on the end plates the electric field is only a radial component in TM mode (see [19]), then $\vec{E}_0 = E_0 e^{i(\omega t - kr\cos\theta)}\vec{u}_r$ in spherical coordinates. One should expect by replacing it in the wave equation (10), the equation for the wavelength [14]

$$k^2 = \mu\epsilon\omega^2 - i\mu\sigma_{Cu}\omega$$
$$k^2 = \mu\epsilon\omega^2(1 - i\frac{\sigma_{Cu}}{\omega\epsilon}) \qquad (11)$$

In the good conductors such as copper, one can make the assumtion [14] that $\frac{\sigma_{Cu}}{\omega\epsilon_0} \gg 1$. Thus, (11) becomes

$$k^2 = \mu\omega(-i\sigma_{Cu})$$
$$k = (1-i)\sqrt{\frac{\sigma_{Cu}\mu\omega}{2}} \qquad (12)$$

Which ends up in an evanescent wave taking into account the real (k_1) and imaginary part (k_2) of the wavelength, $\vec{E} = E_0 e^{-k_1 r\cos\theta} e^{i(\omega t - k_2 r\cos\theta)}\vec{u}_r$. Now, we can estimate the energy density of the EM field $<w> = <u_E> + <u_B>$ with

$$<u_E> = \frac{\epsilon_{Cu}}{2\pi}\int_0^{2\pi} Re\{E.E^*\}dt$$
$$<u_E> = \frac{\epsilon_0}{2\pi}\int_0^{n\tau_r} Re\{E.E^*\}dt$$
$$<u_E> = \frac{\epsilon_0}{2\pi}\int_0^{n\tau_r} E_0^2 e^{-2k_1 r}\cos^2(\omega t - k_2 r\cos\theta)dt$$
$$<u_E> \sim \frac{n\tau_r \epsilon_0}{2\pi} E_0^2 e^{-2k_1 r\cos\theta} \qquad (13)$$

we assume that the Evanescent waves are created by the surface charges only during the relaxation time as explained above. τ_r is part of relaxation time τ_{rel} when the charges create the surface current. In the 2π average interval, there is $n\tau_r$ ($n\tau_r \ll 1$). In the remaining time we consider the integral null. The first derivative of the EM energy density for the electric field is

$$<\partial_t u_E> = \frac{\epsilon_0}{2\pi}\int_0^{2\pi} Re\{2E.\partial_t E^*\}dt$$
$$<\partial_t u_E> = \frac{\epsilon_0 2\omega}{2\pi}\int_0^{n\tau_{rel}} E_0^2 e^{-2k_1 r}\sin(\omega t - k_2 r)\cos(\omega t - k_2 r)dt$$
$$<\partial_t u_E> \sim \frac{n\tau_{rel}\omega\epsilon_0}{2\pi}E_0^2 e^{-2k_1 r}$$
$$<\partial_t u_E> \sim \omega <u_E> \qquad (14)$$

The same development can be applied to the second derivative

$$<\partial_t^2 u_E> \sim 2\omega <\partial_t u_E>$$
$$<\partial_t^2 u_E> \sim \frac{n T_{rel} 2\omega^2 \epsilon_0}{2\pi} E_0^2 e^{-2k_1 r} \quad (15)$$

For the magnetic field, one can estimate with $\vec{curl}\vec{E} = -\partial_t \vec{B}$. Choosing a spherical coordinates referential (Figure),

$$\vec{curl}\vec{E} = \frac{-1}{r}\partial_\theta E \vec{u}_\phi$$
$$= -i\omega \vec{B}$$
$$\vec{B} = (\frac{k_1 \sin\theta}{\omega}(1-i))E\vec{u}_\phi \quad (16)$$

In the same way we estimated $<u_E>$, one can estimate the magnetic energy density

$$<u_B> = \frac{1}{\mu 2\pi}\int_0^{nT_{rel}} Re\{B.B^*\}dt$$
$$<u_B> \sim \frac{1}{\mu\pi}(\frac{k_1 \sin\theta}{\omega})^2 E_0^2 e^{-2k_1 r \cos\theta} n T_{rel}$$
$$<\partial_t u_B> \sim \omega <u_B>$$
$$<\partial_t^2 u_B> \sim 2\omega^2 <u_B> \quad (17)$$

However,

$$\frac{<u_E>}{<u_B>} \sim \frac{\epsilon}{\mu}\frac{\omega^2}{k_1^2 \sin^2\theta}$$
$$\sim \frac{2\epsilon\omega}{\mu^2 \sigma_{Cu}} \gg 1 \quad (18)$$

Because $\frac{<u_E>}{<u_B>} \gg 1$, the energy density of the EM field is mainly the contribution from the electric field. Finally, additional measurements on n can check the assumption on the order of magnitude of the EM energy density.

Simulations and Preliminary Results

In this section, simulations of the copper frustum in TM010 mode has been performed by Christian Ziep using FEKO software. The frustum is model as described in [5] and [22] without a dielectric insert. It is orientated following the Z-axis with the direction pointing towards the small end plate. The dimension of the cavity follows: 228.6 mm (height),

158.75 mm (diameter small end plate), 279.65 mm (diameter big end plate). The antenna model is an electrical dipole placed in the middle of the cavity. The input power is equal to 1W (30 dBm) with central frequency 0.9598 GHz and quality factor Q equal to 20.38. The resonant frequency is then estimated at 1020 MHz. Figure 2(A) displays the magnetic field inside the cavity perpendicular to the conical wall and parallel to the end plates as described in [18] and [19]. Figure 2(B) displays the electric field perpendicular to the end plates.

Now, the surface currents on the cavity walls are simulated following the previous description. Figure 3 (A,B) display the amplitude of the electric (E) and magnetic fields (H) at the surface of the conical wall as a function of the height; Figure 3 (C,D) the amplitude of the E and H-field at the surface of the small end plate; and Figure 3 (E,F) the amplitude of the E and H fields at the surface of the large end plate. The results show that the amplitude of the currents at the surface of the conical wall follows a gradient decreasing with the increase of the height of the frustum. It is in agreement with the observations that both E and H fields are larger (on average) at the surface of the large end than at the small end. Thus, the gradient of the amplitude of the wall current accommodates with the amplitude of simulated E and H fields at the surface of the end plates.

One assumption in our MLT model is the current propagating from large to small end plate due to the difference of electrical potential. In the simulations, the current at the surface of the conical wall propagates towards the large end plate. Thus, it seems that those currents are Eddy currents generated by the H field. As previously underlined, the Eddy currents could have higher amplitude than the one due to difference of electrical potential. This result underlines this phenomenon. In addition, the electric field at the surface of the large end plate is higher than at the small end plate, which supports a greater EM excitation. Based on our assumption that the Woodward effect is directly related to the skin depth effect taking place on the cavity wall, this effect should then be greater on the large end than on the small end. It has been shown in [6] that, in this experiment, the anomalous thrust is towards the large end plate. This result is in agreement with (9), assuming that the Woodward effect displaces the cavity towards the large end due to $\delta\rho_L > \delta\rho_S$. However, further study is required to understand the role of the Lorentz force taking place on the conical wall in the amplitude of the anomalous thrust.

3. Concluding Remarks

This model was based on a few results on the TM010 mode (i.e. [5] and [6]) and preliminary simulations. The study takes into account the EM excitation of each element of the cavity resulting in modelling them with a capacitor with a resistance in series, and an inductor in parallel. Thus, two types of currents are then taking into account: Eddy currents

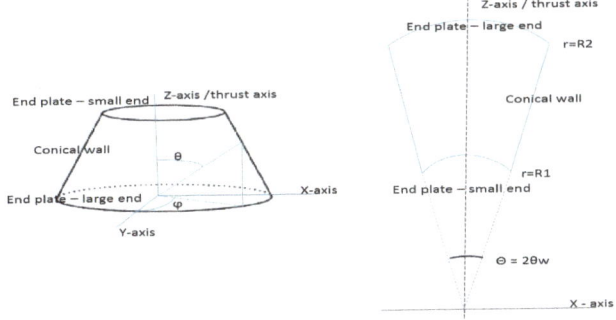

Figure 1: Drawing of the EM Drive cavity

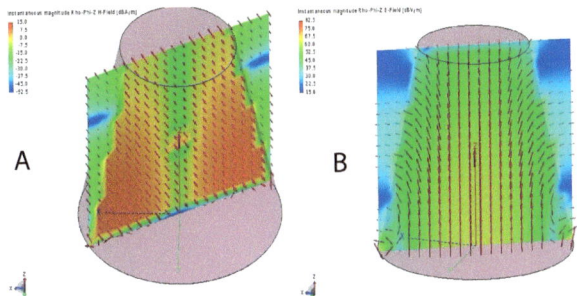

Figure 2: Simulations of the EM field inside the frustum in TM010 mode: (A) magnetic field, (B) electric field

from transverse magnetic field and surface currents from electric field excitation. It is then produced a surface currents (dI_0) on the conical wall, hence creating a Lorentz force. The last step of our model is the generation of thrust using the Woodward effect. However, the thrust is only produced by a coupling between the Lorentz force and the Woodward effect from (9) in order to guarantee momentum conservation principle. Only a careful analysis via simulations and experiments of the frustum for a specific mode can quantify the contribution of those currents to the proposed model of the thrust.

The proposed model is just at an early development stage where many assumptions must be validated. For example, the theory stands at the moment with those few points to check:

- Estimation of the currents on the cavity walls due to the electric and magnetic fields.

- On the need to estimate the AC current I_0 on the conical wall and the Lorentz force $\vec{F}_{Lo}$ through simulations and experiments with different scenarios (e.g., with and without HDPE).

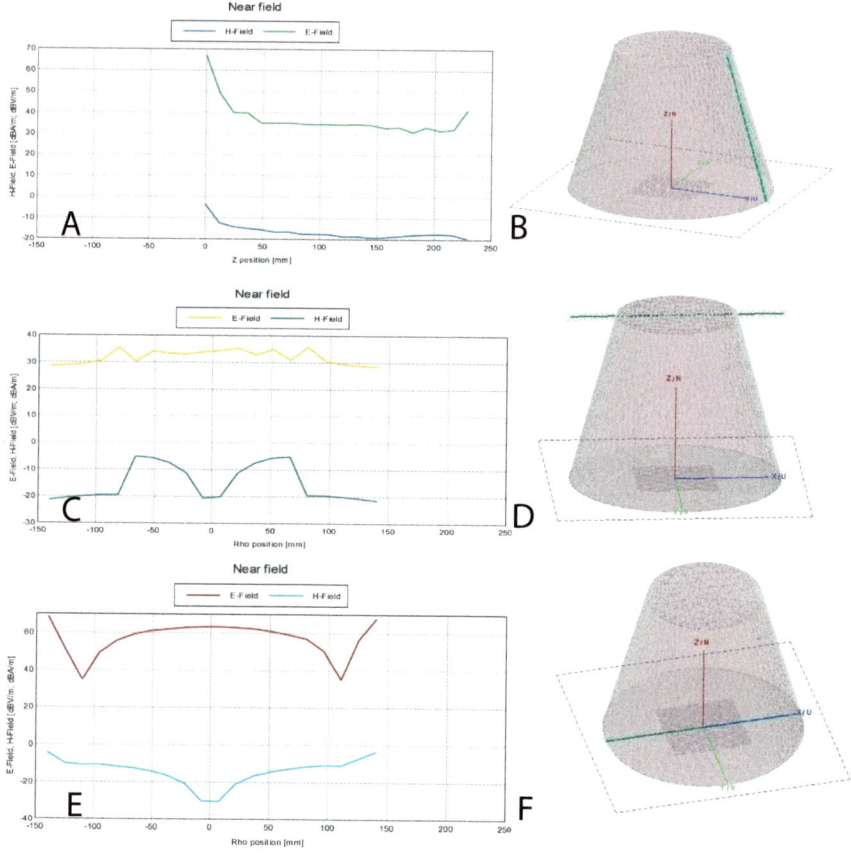

Figure 3: Estimation of surface currents (A,B) conical wall, (C,D) small end, (E,F) large end

- Better understanding of the coupling between the acceleration of the cavity due to $\vec{F}_{Lo}$ and the Woodward effect.

- The variation of mass $\delta\rho(t)$ in (8) with the first and second derivatives of the EM energy density.

Overall, our assumptions on this model have to be compared with the results from following experimentations. One can underline

- The model can be invalidated if there is still a non negligible thrust if we use superconductive materials for the frustum in order to eliminate (or reduce drastically) all skin depth effects on the cavity walls (suggested by Prof. J. Woodward).

- The first and second steps of this model rely on standard EM theory. One needs to estimate the average electric field at the surface of the end plates in order to get some measurements for the amplitude of the difference of electric potential (DV) and also to confirm the simulations.

Furthermore, at the time of writing this manuscript, NASA Eagleworks laboratory has released a full study supporting the EM Drive generating a thrust in TM mode [5,6]. New experiments are planned to test the TE mode, which can help supporting or not this model. We also acknowledge that some engineers have recently carried out tests involving various new designs of the EM Drive showing successfully an anomalous thrust. The TE mode is the next step in order to produce a complete MLT model of the EM Drive and its anomalous thrust. To conclude, this new engine can be an example of EMG coupling if the presented model is validated.

4. Acknowledgements

The author would like to acknowledge people who have also been involved in developing the presented model via discussions or various feedbacks including Paul March (NASA, Eagleworks), Todd J. Desiato (Warp Drive Tech.) and Christian Ziep. Christian did some preliminary simulations of the EM Drive cavity using FEKO software.

References

[1] R. Shawyer, *The EM Drive a New Satellite Propulsion Technology*, in Proc. of the 2nd Conference on Disruptive Technology in Space Activities, 2010.

[2] R. Shawyer, *Second generation EmDrive propulsion applied to SSTO launcher and interstellar probe*, Acta Astronautica, p. 166-174. Doi:$http://dx.doi.org/10.1016/j.actaastro.2015.07.002$

[3] H. Fearn, A. Zachar, J. F. Woodward and K. Wanser, *Theory of a Mach Effect Thruster*, in Proc. of the AIAA Joint Propulsion

Conference, Tech. Session: Nuclear and Future Flight Propulsion, $http://arc.aiaa.org/doi/abs/10.2514/6.2014 - 3821$.

[4] J. F. Woodward, *Life Imitating Art: Flux Capacitors, Mach Effects, and Our Future in Spacetime*, AIP Conference Proceedings. Space Technology Applications International Forum (STAIF 2004), Albuquerque, New Mexico. American Institute of Physics. p. 11271137, 2004. Doi: 10.1063/1.1649682.

[5] D.A. Brady, H.G. White, P. March, J.T. Lawrence, and F.J. Davies, *Anomalous Thrust Production from an RF Test Device Measured on a Low-Thrust Torsion Pendulum*, AIAA 2014-4029, 2014.

[6] H. White, P. March, J. Lawrence, J. Vera, A. Sylvester, D. Brady, P. Bailey, *Measurement of Impulsive Thrust from Closed Radio Frequency Cavity in Vacuum*, AIAA Journal of Propulsion and Power, December, 2016.

[7] M. Tajmar, and G. Fiedler, *Direct Thrust Measurements of an EM Drive and Evaluation of Possible Side-Effects*, in Proc. of the 51st AIAA /SAE /ASEE Joint Propulsion Conference, AIAA 2015-4083. Doi: $10.2514/6.2015 - 4083$

[8] J.F. Woodward, *Making Starships and Stargates*, Springer-Verlag New York, 2013. Doi: $10.1007/978 - 1 - 4614 - 5623 - 0$

[9] M. E. McCulloch, *Can the Emdrive Be Explained by Quantised Inertia ?*, Progress in Physics, vol. 11, 2015.

[10] P. Grahn, A. Annila, E. Kolehmainen, *On the Exhaust of Electromagnetic Drive*, AIP Advances, vol. 6 (6), 2016. Doi: 10.1063/1.4953807

[11] T. J. Desiato, *An Engineering Model of Quantum Gravity*, September, 2016.

[12] J.F. Woodward, *Gravity, Inertia, and Quantum Vacuum Zero Point Fields*, Foundations of Physics, 31 (5), p. 819835, 2001. Doi: 10.1023/A:1017500513005

[13] D. W. Sciama, *On the Origin of Inertia*, Monthly Notices of the Royal Astronomical Society, vol. 113 (1), p. 34-42, 1953. Doi: $10.1093/mnras/113.1.34$

[14] R. Petit, *Ondes Electromagnetiques en radioelectricite et en optique*, 2nd Edition, Masson, 1993.

[15] A. E. Fitzgerald, C. (Jr) Kingsley, S. D. Umans, Electric Machinery, 4th ed., Mc-Graw-Hill, Inc. p. 20, 1983. (ISBN 0-07-021145-0)

[16] G.G. Raju, *Dielectrics in Electric Fields*, CRC Press, 2003. (ISBN:9780824708641).

[17] MIT Web lecture, *Resonant Cavities and Waveguides*, http://web.mit.edu/22.09/ClassHandouts/ Charged Particle Accel/CHAP12.PDF

[18] J. Rodal, *Resonant Cavity Space Propulsion* `http://forum.nasaspaceflight.com`

[19] G. Egan, *Resonant Modes of a Conical Cavity*. http://gregegan.customer.netspace.net.au/SCIENCE/Cavity/Cavity.html

[20] David R. Lide, *CRC Handbook of Chemistry and Physics*, CRC Press Inc, 2009, 90e Ed., 2804.(ISBN 978-1-420-09084-0)

[21] D. H. Staelin, *Electromagnetics and Applications*, Department of Electrical Engineering and Computer Science, Massachusetts Institute of Technology, Cambridge MA. Available at: `https://ocw.mit.edu/courses/` or google MIT6_013S09_notes.pdf

[22] F. J. Davies, *Copper Frustum Modes*, Personal Communication, NASA Johnson Space Center, NASA/JSC/EP5.

[23] J.P. Montillet, *The Generalization of the Decomposition of Functions by Energy Operators (Part II) and Some Applications*, Acta Applicandae Mathematicae. doi: 10.1007/s10440-014-9978-9, also available at: `http://arxiv.org/abs/1308.0874`.

[24] J.P. Montillet, *Multiplicity of Solutions for Linear Partial Differential Equations Using (Generalized) Energy Operators*, available in: `http://arxiv.org/pdf/1509.02603v1.pdf`

[25] E. B. Porcelli, V. S. Filho, *On the Anomalous Weight Losses of High Voltage Symmetrical Capacitors*, ArXiv. doi: 10.4006/0836-1398-29.1.002

[26] C. Möller, *The Theory of Relativity*, 2nd ed., Delhi: Oxford University Press. p. 220, 1952. ISBN 0-19-560539-X.

Editor's Note: Addendum I and Addendum II, on a new mathematical framework, are omitted from this focus volume. Those Addenda can be found in the James F. Woodward commemorative proceedings on the Space Studies Institute website.

Addendum III: Derivation of the Woodward effect using the electromagnetic energy density

Assumptions with the energy momentum relationship

When the Woodward effect was established in [4], the authors implicitly assumed the rest mass of the piezoelectric material via the famous Einstein's relation in special relativity $\mathcal{E} = mc^2$ ($\mathcal{E}$ the rest energy associated with the rest mass m) and its variation via electrostrictive effect.

Here, the system is the frustum. The rest mass is all the particles within it at the time of the capacitor is discharged. It excludes the photons considered with a null mass. Thus, the main assumption is that the EM excitation on the end plates creates electric charges (i.e. electrons) which makes the rest mass varying with time. This assumption is the same as the mass variation of a capacitor between the charge and discharge times [25]. It allows us to state the variation of rest energy such as:

$$\begin{aligned} \Delta \mathcal{E} &= \mathcal{E}(t+dt) - \mathcal{E}(t) \\ &= (m(t+dt) - m(t))c^2 \\ &= \Delta m c^2 \end{aligned} \quad (19)$$

Finally, the variation of rest energy $\Delta\mathcal{E}$ is assumed to be equal to the variation of EM energy density (Δu_{EM}) resulting from the charges within the skin depth of the copper walls. We also cannot forget the electrostrictive effect (Δu_{El}) when inserting HDPE disk(s) inside the frustum, but we consider that $\Delta u_{EM} >> \Delta u_{El}$.

Note that at the particle level, the rest mass should satisfy the energy momentum relationship for a free body in special relativity [26]:

$$\begin{aligned} u_e^2 &= (pc)^2 + (m_e c^2)^2 \\ p &= v \frac{u_e}{c^2} \end{aligned} \quad (20)$$

with p the momentum and m_e the rest mass of the particle associated with the total energy u_e. The particle is accelerated via the Lorentz force applied to the whole cavity with obviously $v << c$. Thus, we have also the relationship $p^2 < (u_e/c)^2$.

Woodward effect

From [4], one can write the mass variation per unit of volume

$$\begin{aligned} dm &= \frac{\delta m}{V} \\ dm &= \frac{1}{4\pi G}[\frac{1}{m}\partial_t^2 m - \frac{1}{m^2}(\partial_t m)^2] \end{aligned} \quad (21)$$

If we define the mass density such as $\rho = m/V$, then

$$\delta\rho = \frac{\delta m}{V}$$
$$\delta\rho = \frac{1}{4\pi G}[\frac{1}{\rho}\partial_t^2\rho - \frac{1}{\rho^2}(\partial_t\rho)^2] \tag{22}$$

Let us define the the rest energy $\mathcal{E} = \rho c^2$, then

$$\delta\rho = \frac{1}{4\pi G}[\frac{1}{\rho c^2}\partial_t^2\mathcal{E} - \frac{1}{(\rho c^2)^2}(\partial_t\mathcal{E})^2]$$
$$\delta\rho = \frac{1}{4\pi G}[\frac{1}{\mathcal{E}}\partial_t^2\mathcal{E} - \frac{1}{(\mathcal{E})^2}(\partial_t\mathcal{E})^2] \tag{23}$$

Now, with the assumption that the variation in time of the rest energy is equal to the variation of EM energy density u

$$\delta\rho = \frac{1}{4\pi G}[\frac{1}{\mathcal{E}}\partial_t^2 u - \frac{1}{(\mathcal{E})^2}(\partial_t u)^2] \tag{24}$$

The EM energy density u follows the general definition of the sum of energy density from the electric (u_E) and magnetic (u_B) fields [14].

✔ Fuel Problem	✔ Theory	Existing Physics
✔ Time-Distance Problem	Experiment	✔ New Physics

Kaluza Unification of Gravity and EM

- Lance Williams will lead a discussion about the Kaluza theory as a promising framework to address the fuel and time-distance problems.

Issue Summary:

Lance will present the Kaluza unified theory of gravity and electromagnetism. It was developed in the few years between the publication of GR, and the quantum revolution of Heisenberg and Schroëdinger. It provided a compelling unification of gravity and electromagnetism that accorded with the existing theory in every way. Einsten himself spent several years working on it, and it has formed the basis of subsequent higher-dimensional quantum theories.

The Wikipedia page on Kaluza-Klein theory provides a very good summary of the Kaluza theory. Lance has a recent article [2] that derives the full Kaluza equations for the first time, using tensor algebra software. The Lagrangian is also established.

Essentially, Kaluza found the the equations of GR and EM could be recovered by writing the equations of GR in 5 dimensions. The 15 components of the 5D metric are now accounted for by the 10 components of the 4D metric of GR, the 4 components of the electromagnetic four-vector potential, and an unidentified scalar field. The Einstein equations in 5D produce the 4D Einstein equations, the Maxwell equations, and an equation for the scalar field. Simultaneously, and independently, the 5D geodesic equation produces the 4D geodesic equation and the Lorentz force law of EM. Remarkably, the 5D theory written in vacuum yields the 4D EM stress energy tensor in the Einstein equations: an example of "matter from geometry".

The theory promises a solution to the fuel problem through an adjustable coupling of gravity, and a solution to the time-distance problem through a hyper-dimension. Both of these are through electromagnetic means. So far, no experimental verification has been suggested that would verify the theory. Lance hopes to present one to the group for discussion.

Objectives of the discussion will be to understand the motivation behind the Kaluza theory; and to address experimental ways to validate the theory.

1. https://en.wikipedia.org/wiki/Kaluza-Klein_theory
2. L.L. Williams, Journ. of Gravity, v2015, ID 901870, http://dx.doi.org/10.1155/2015/901870

Estes Park Advanced Propulsion Workshop	Block 3	20 September 2016

Experimental Verification of the Kaluza Theory of Gravity and Electromagnetism

L. L. Williams
Konfluence Research Institute
Manitou Springs, Colorado, USA

The Kaluza unified theory of general relativity and Maxwellian electrodynamics has intrigued researchers for a century. However, no new prediction of the theory, not already found in either existing general relativity or electrodynamics, has ever been proposed – until now.

Estes Park Categories

According to the categories established in the prep kit for this session, this is a mature theory that completely includes general relativity and classical electromagnetism. The theory is relevant to the time-distance problem, as summarized in the pre-workshop Prep Kit. No experiment to falsify the theory has been proposed until now.

1. Motivation: A Syllogism

We are driven by an irresistable logic to search for a solution to the time-distance problem within the framework of the Kaluza theory.

1. The time-distance problem is one of special and general relativity. (cf. Estes Park Quick Study II)

2. Electromagnetism is the only force of nature humans control. Our machinery, our communications, our power generation, even our chemistry and metallurgy – is all electromagnetic

3. The human solution to the time-distance problem may be found at the intersection of general relativity and electromagnetism.

4. The Kaluza unification of general relativity and electromagnetism is the only theory to sit at that intersection, and is the only unified theory ever found that encompasses general relativity

2. Kaluza, not Klein

The Kaluza theory under consideration here is purely classical. Kaluza submitted his paper to Einstein in 1919 [2], and Einstein forwarded it for publication in 1921 [1]. In 1925 came the quantum revolution of Schroedinger and Heisenberg. A search was soon undertaken to find quantum versions of the classical field theories. Electrodynamics was successfully quantized in the 1940s, but general relativity never was. In 1926, Klein entered the picture by translating Kaluza's purely classical picture into a quantum interpretation. [3] Klein hypothesized that Kaluza's fifth dimension was closed and microscopic. The term "Kaluza-Klein" has been used ever since, but I emphasize we consider only Kaluza here, not Klein. We presume Kaluza's fifth dimension to be macroscopic, as he originally understood it, not closed and microscopic, as Klein understood it. Also note that a series of researchers working independently [4] have worked on the field equations to be summarized below.

3. The Power of the Einstein Equations

The basic hypothesis of Kaluza is that the Einstein equations hold true in five dimensions, not just the familiar four of space and time. Why should we extend Einstein's framework to five dimensions? In fact, there are good reasons for doing so.

The Einstein equations are the most general covariant, second-order differential equations, that can be written. In effect, they generalize Newton's law of gravity $\nabla^2 \phi = 4\pi G \rho$. They have a supreme mathematical power that is not particular to the number of dimensions in which they are written. In fact, after Kaluza, other workers began to pursue unified field theories by writing the Einstein equations in even higher numbers of dimensions.

The standard Einstein equations are typically written

$$R_{\mu\nu} - \frac{1}{2}g_{\mu\nu}R = \frac{8\pi G}{c^4}T_{\mu\nu} \qquad (1)$$

The equivalent Lagrangian, discovered by Hilbert, is

$$\mathscr{L} = g^{1/2}\left(\frac{c^4}{16\pi G}g^{\alpha\beta}R_{\alpha\beta} + T_{\mu\nu}g^{\mu\nu}\right) \qquad (2)$$

where the quantities are as described in my earlier talk on aspects of a valid extension to the laws of physics. But now, the factor $g^{1/2}$, necessary to preserve an invariant volume element, is shown explicitly. The action $S = \int \mathscr{L} d^4x$. Where Newton identified a scalar gravitational potential, Einstein identifies a symmetric metric tensor $g_{\mu\nu}$ with 10 components. Note also the unique coupling of the metric, in that it enters both terms of (2), even as $R_{\mu\nu}$ is composed of derivatives of the metric.

4. Kaluza Hypothesis: Vacuum Field Equations

Although Kaluza originally considered a five-dimensional source for the equations, we consider here the five-dimensional (5D) field equations in vacuum.

The 5D vacuum field equations:

$$\tilde{R}_{ab} - \frac{1}{2}\tilde{g}_{ab}\tilde{R} = 0 \tag{3}$$

where roman indices span the five dimensions, and a tilde is used to indicate 5D tensors.

The Lagrangian for these field equations was verified using tensor algebra software [5] to be the standard Hilbert Lagrangian:

$$\mathscr{L} = \tilde{g}^{1/2}\tilde{R} \tag{4}$$

In addition, Kaluza applied the constraint known as the cylinder condition. It is that no field variable depends on the fifth coordinate:

$$\frac{\partial \tilde{g}_{ab}}{\partial x^5} = 0 \tag{5}$$

This is the mathematical expression of the fact that the fifth dimension is not detected: there is no variation in that coordinate. Klein alternatively "explained" why we don't detect a fifth dimension by hypothesizing the fifth coordinate was compact and microscopic.

If derivatives with respect to the fifth coordinate are included, the resulting field equations yield many more terms. Since standard general relativity and electromagnetism are recovered under the cylinder condition, some researchers have relaxed the cylinder condition, and identified the resulting extra terms with an effective stress-energy tensor; viz., matter arises from geometry.

The 15 components of the 5D metric $\tilde{g}_{ab}$ are decomposed into the 10 components of the 4D metric $g_{\mu\nu}$, the 4 components of the electromagnetic vector potential A^μ (the constant k preserves units), and a scalar field ϕ.

$$\begin{aligned}\tilde{g}_{\mu\nu} &= g_{\mu\nu} + \phi^2 k^2 A_\mu A_\nu, \quad \tilde{g}_{\mu 5} = \phi^2 k A_\mu, \quad \tilde{g}_{55} = \phi^2 \\ \tilde{g}^{\mu\nu} &= g^{\mu\nu}, \quad \tilde{g}^{5\mu} = -kA^\mu, \quad \tilde{g}^{55} = A_\mu A^\mu + 1/\phi^2\end{aligned} \tag{6}$$

Greek indices span the 4 dimensions of space and time. Kaluza originally set $\phi = 1$, and ignored its dynamics. Subsequent researchers such as Jordan, Thiry, Brans, and Dicke, have investigated this version of scalar field theories and others [4].

5. Vacuum Field Equations: Coupling of Gravity and Electromagnetism

In a recent work, using tensor algebra software, I was able to show [5] that the vacuum Lagrangian under the Kaluza hypothesis is

$$\mathscr{L} = g^{1/2}\left(\frac{c^4}{16\pi G}\phi g^{\alpha\beta}R_{\alpha\beta} - \frac{1}{4\mu_0}\phi^3 g^{\alpha\mu}g^{\beta\nu}F_{\alpha\beta}F_{\mu\nu}\right) \qquad (7)$$

This is quite an unusual Lagrangian, because the scalar field enters only algebraically. Nearly all researchers have been tempted to insert by hand a term proportional to $\partial_\alpha \phi \partial^\alpha \phi$. Yet it is inescapable that the Hilbert Lagrangian (4) and the decomposition (6) lead to the Lagrangian (7).

To consider the field equations from this decomposition, we evaluate the components of $\widetilde{G}_{ab}$ as $G_{\mu\nu}$, $G_{5\mu}$, and G_{55}.

$$\widetilde{G}_{\mu\nu} = 0 \quad \longrightarrow \quad R_{\mu\nu} - \frac{1}{2}g_{\mu\nu}R = \frac{8\pi G}{c^4 \mu_0}\phi^2 T^{EM}_{\mu\nu} + T^\phi_{\mu\nu} \qquad (8)$$

where

$$T^{EM}_{\mu\nu} \equiv g^{\alpha\beta}F_{\mu\alpha}F_{\nu\beta} - \frac{1}{4}g_{\mu\nu}F_{\alpha\beta}F^{\alpha\beta} \quad , \quad T^\phi_{\mu\nu} \equiv \frac{1}{\phi}(\nabla_\mu \partial_\nu \phi - g_{\mu\nu}\Box\phi)$$

The result (8) is quite extraordinary. From a vacuum equation, we obtained sources to the 4D field equations. It is extraordinary how the correct form of the electromagnetic stress-energy tensor falls out, and this has been called the "Kaluza miracle" that makes the theory so attractive. It is also extraordinary that we get a conventional stress-energy tensor for the scalar field, even though the scalar field enters only algebraically in the Lagrangian (7). Also note that (8) allows the constant k to be fixed in terms of the gravitational constant and the permeability of free space.

Of particular interest for the propulsion problem is that the scalar field enters as a variable coupling between gravity and electromagnetic energy. It would effectively lead to a variable gravitational constant.

Moving on to another aspect of the Kaluza miracle,

$$\widetilde{G}_{5\nu} = 0 \quad \longrightarrow \quad \nabla^\mu(\phi^3 F_{\mu\nu}) = 0 \qquad (9)$$

Here are the covariant vacuum Maxwell equations, but now with an effective displacement-current type source in the scalar field.

Finally,

$$\widetilde{G}_{55} = 0 \quad \longrightarrow \quad \frac{12\pi G}{c^4 \mu_0}\phi^2 F_{\alpha\beta}F^{\alpha\beta} = R \qquad (10)$$

Again, as can be verified trivially from the Lagrangian (7), the equation for the scalar field is algebraic with no dynamics.

Question: The expression for the metric (6) involves also the quantity A^2. Where is the equation for that?

Williams: It is accounted in the equations for $F_{\alpha\beta} \equiv \partial_\alpha A_\beta - \partial_\beta A_\alpha$. We do have 15 equations in 15 unknowns and everything is accounted for.

6. Equations of Motion: 5D Geodesic Equation

While the field equations describe how the fields respond to sources, the equations of motion describe how matter responds to the fields. The geodesic equation in 5D is:

$$\frac{d\widetilde{U}^c}{d\theta} + \widetilde{\Gamma}^c_{ab}\widetilde{U}^a\widetilde{U}^b = 0 \quad , \quad \widetilde{U}^a \equiv \frac{dx^a}{d\theta} \tag{11}$$

Decomposing this into U^μ and U^5, we find

$$\frac{dU^\mu}{d\tau} + 2\widetilde{\Gamma}^\mu_{5\alpha}U^\alpha U^5 + \widetilde{\Gamma}^\mu_{55}(U^5)^2 + U^\mu \frac{d}{d\tau}\ln\left(\frac{d\tau}{d\theta}\right) = 0 \tag{12}$$

Now, the motion of a body under combined gravitational and electromagnetic forces is known to be

$$\frac{dU^\mu}{d\tau} + \Gamma^\mu_{\alpha\beta}U^\alpha U^\beta = \frac{Q}{m}F^\mu{}_\nu U^\nu \tag{13}$$

Another Kaluza miracle is that $\widetilde{\Gamma}^\mu_{5\nu} \propto F^\mu{}_\nu$, inviting us to identify U^5 with the charge-to-mass ratio Q/m. The term in U^5 caused some concern at first, because U^5 can be large for elementary particles, and then this term would dominate the equations of motion. However, $\widetilde{\Gamma}^\mu_{55} \propto \partial_\alpha \phi$, and so this term vanishes in regions of constant ϕ.

7. Charge = Motion

Using the value for k in (6) fixed by (8), we can relate:

$$\frac{dx^5}{d\tau} \to c\frac{Q/m}{\sqrt{16\pi G\epsilon_0}} \tag{14}$$

Mathes: Are you saying the fifth dimension stores charge? Is it stored in the compact fifth dimension? Is this related to quantum theory or will it be reconciled with quantum theory?

Williams: First, this is a purely classical theory. There is to be no reconciliation with quantum theory, and Planck's constant does not enter. A century ago, researchers were not interested in this theory because there seemed to be no point at trying to unify gravity and electromagnetism at the classical level, since we know they are quantum fields. But since then, no quantum theory of gravity has been found. The century's greatest minds – Feynman, Schwinger, Paul, Dirac – were unable to solve that riddle. How long should we keep trying? As to your other question, the fifth dimension does not store charge. Rather, what we call charge is actually due to motion in the fifth dimension. When you see a charged particle at rest, it is actually traveling through the fifth dimension while its spatial coordinates remain unchanged.

We see that electric charge is a fifth component of an energy-momentum-charge 5-vector:

$$\widetilde{U}^a = \frac{\gamma_5}{m}(E/c, \mathbf{p}, Q/k) \tag{15}$$

As energy is due to motion in time, and momentum is due to motion in space, so electric charge is due to motion in the fifth dimension.

Buldrini: This seems to imply that the speed with which the electron moves in the fifth dimension is absolutely constant and unchanging.

Williams: Yes, that's right. In fact, speeds are astronomical for elementary particles, $\sim 10^{20}c$.

Hansen: Earlier you said the cylinder condition stipulated no change in the fifth dimension, so how can things be moving in the fifth dimension?

Williams: Those are two different things. The cylinder condition tells us no fields depend on the fifth coordinate, but that does not mean things are not moving (uniformly) through the fifth dimension.

Mathes: Can you remind us of the definition of proper time?

Williams: Yes, it is the invariant length element. In 4D, $d\tau^2 = c^2 dt^2 - dx^2$. In 5D, we add in a piece such that $d\theta^2 = c^2 dt^2 - dx^2 - d(x^5)^2$.

8. Time Dilation of Charged Clocks

Here we propose a way to verify or falsify the Kaluza hypothesis. As just mentioned, we can write the 5D invariant length element at a fixed position ($\Delta x = 0$):

$$d\theta^2 = c^2 dt^2 + d(x^5)^2 = c^2 dt^2 (1 + \beta_5^2) \quad ; \quad \beta_5^2 \equiv \frac{Q^2/m^2}{16\pi G\epsilon_0} \tag{16}$$

We see here the core features of a time dilation or time contraction effect. If there is a rest frame in the fifth dimension, then a clock could feasibly run faster or slower, depending on its motion in the fifth dimension. It is analogous to the time dilation exhibited by muons produced in the upper atmosphere. But instead of motion in space, we wish to consider motion in the fifth dimension. These considerations raise several questions.

Is there a rest frame in the fifth dimension?

A major distinction between space and time is that the former allows a rest frame for massive objects, whereas the latter does not. No massive object can be at rest in time, but one can be at rest in space. If the fifth dimension has no rest frame, then this effect may not be detectable. But if the fifth dimension is spacelike in that it does allow a rest frame, then time dilation or time contraction may be something we can find a way to detect.

What is the meaning of the mass of a clock?

The mapping of speed to charge (14) requires a mass. Considering clocks in spatial motion, we do not encounter this conceptual issue. But in this case, we wish to consider a rest frame characterized by a charge-to-mass ratio.

If there were a rest frame to the fifth dimension, and if the effect were detected, it would be a new effect unknown to physics.

10. Properties of a Charged Clock

The first property of a charged clock is that it must be uniformly charged throughout its volume. Under the Kaluza hypothesis, only charged particles are moving through the fifth dimension. But if those particles merely surround a volume that is itself uncharged, then the interior volume is not moving in the fifth dimension, and is not charged in the sense implied here.

Volume charges are perhaps more difficult to realize than surface charges, so perhaps a two-dimensional clock, if it could be found, could be more easily charged.

> **Meholic:** Perhaps the effect could be somehow realized with an atomic clock such as a cesium clock.
>
> **Williams:** Yes, I had also thought of using the decays of nuclides somehow.
>
> **Cole:** Could it be realized with a capacitor?

Meholic: It seems plausible because the discharge time might depend on the charges on the capacitor plate.

Buldrini: Perhaps an electret.

March: Are you familiar with Erwin Saxl's experiments with electrified pendulums in 1964? He found some strange time effects. It was in Nature, 1964. Earl Saxl.

Tajmar: Before you repeat that experiment you should know it was found out to be due to electrostatic charge on the chamber walls.

Another key property of a charged clock is that it must offer two different charge-to-mass states, but without affecting its timekeeping mechanism or principle. This is because we need to test how time varies depending on the charge state.

Broyles: Could you use a double slit experiment, using charged bucky balls? Do it charged and uncharged. Evaluate the time of flight to the detector.

Cole: I would come back to the capacitor. Except, use just one plate. The plate is charged and can be considered a 2D clock. The clock is the discharge time of the capacitor.

Tajmar: You must consider that if you use alternating currents, you can get accelerations and radiation, and makes it difficult to control. Ideally, you will have linear motion with no acceleration and no radiative losses. I would approach this with a mass spectrometer and time of flight considerations. You can control the charge state of the ions, and measure the time of flight very accurately.

Williams: I don't understand why time of flight would be a charged clock. I think it's motion in space is not really the motion we are trying to isolate in the fifth dimension.

Tajmar: Perhaps you could charge up a macroscopic quantity of something radioactive, and see if its decay rate changes. Perhaps a cesium plasma would constitute a 3D charge state.

Buldrini: Aren't any of these schemes just moving charges around? The charges don't change, whatever material they may be in. It is not clear what is decaying.

Moving along, let us consider the alternative case, where there is no rest frame to the fifth dimension. In that case, the time coordinate and the fifth coordinate are bound together for charged particles. This situation holds promise in that there may be an effective speed c' and an effective time t' such that $(c')^2(dt')^2 = c^2dt^2 + (dx^5)^2$. Since the coefficient of the time coordinate is what sets the limiting speed (speed of light – cf. Estes Park Quick Study II), perhaps this situation would promise an alternative, effective limiting speed that could be manipulated with electric charge.

11. Time Dilation or Contraction from Electric Charge, in Existing Physics

The most famous example of time dilation about an electrically charged region is given by the Reissner-Nordstron metric. The time component is:

$$d\tau^2 = dt^2 \left(1 - \frac{2GM}{rc^2} + \frac{Q^2 G}{r^2 4\pi\epsilon_0 c^4}\right) \quad (17)$$

Although the mass M causes the familiar gravitational time dilation, the electric charge Q causes a counter-acting time contraction. Note that the variation in time is proportional to the gravitational constant, and independent of any mass.

However, the Reissner-Nordstrom metric applies to the vacuum surrounding a charged mass. We are interested in the interior, where the charge is. Let us turn to the interior solution result of Arnowitt et al [6]. They find the balance between mass-energy, gravitational energy, and electrostatic energy, is given by:

$$mc^2 = m_0 c^2 + \frac{Q^2}{2r}\frac{1}{4\pi\epsilon_0} - \frac{GM^2}{2r} \quad (18)$$

In the limit that $r \to 0$, the mass does not vanish, but acquires a value

$$M = \pm Q/\sqrt{4\pi\epsilon_o G} \quad (19)$$

In this case, we have a scaling of $M^2 \sim Q^2/G$. This is the same scaling as (14). So we see that the scaling from the Kaluza result is consistent with expectations from existing classical theory.

12. Summary of Breakthrough Potential

Let us summarize the aspects of the Kaluza theory that hold promise for a breakthrough in propulsion.

- An electromagnetically-tuneable coupling of mass to gravity
- A hyperspace dimension with large characteristic speeds

- Electromagnetic control of the flow of time, if the fifth dimension has a rest frame

- A way to re-scale the time-distance problem, if the fifth dimension has no rest frame

Woodward: I am not sure you will be able to tune the coupling of gravity, because it is not clear what the charge-carriers are for the scalar field. To control electromagnetism we move charges and currents around. I am not sure this will be workable unless you have scalar charge carriers that you can manipulate.

Williams: Your point is well taken. Kaluza did write down the source terms for the scalar field, and it is something like "charge squared", the 5-5 component of the stress-energy tensor. But I don't know what that means physically, or how to isolate the scalar charges.

P. Jansson: What about an excited atom? Would the decay time depend on its energy?

References

[1] T. Kaluza, *Sitzungsberichte der K. Preussischen Akademie der Wissenschaften zu Berlin*, 966 (1921).

[2] A. Pais, *Subtle is the Lord: The Science and the Life of Albert Einstein*, Oxford University Press: Oxford (1982). (cf. section 17c)

[3] O. Klein, *Z. Phys.*, **37**, 76 (1926).

[4] H. Goenner, *Gen. Relativ. Gravit.*, **44**, 2077 (2012).

[5] L. Williams, *Journal of Gravity*, Article ID 908170 (2015), http://dx.doi.org/10.1155/2015/901870

[6] R. Arnowitt, S. Deser, & C. Misner, *Physical Review*, 120 (1960).

✔ Fuel Problem	✔ Theory	Existing Physics
✔ Time-Distance Problem	Experiment	✔ New Physics

Modified Chameleon Density Model

◆ Tony Robertson will lead a discussion about a propulsion application of the Chameleon Cosmology model

Issue Summary:

The Chameleon Cosmology theory was proposed by Khoury and Weltman in 2004; see [1] and [2]. A useful overview is [3]. The theory involves a scalar field with a variable coupling to matter, proposed to explain Dark Energy by providing a reduction to gravity. This theory is in the family of scalar field theories whose most famous member is the Brans-Dicke theory of gravity, which augmented GR with a scalar field that depended on the mass density of the universe, and identified that scalar field with the gravitational constant. In this case, the coupling is apparently variable. The Wikipedia page [4] is short, but has additional references and detail on experimental implications.

Tony explores the implications for acceleration. He will discuss the thin-shell aspect of Chameleon Cosmology, and point out connections to the Alcubierre bubble. Finding a means for propellantless propulsion with this framework, Tony will also touch on application to seemingly-unrelated thrust experiments, such as the RF resonant cavity, the Podkletnov superconductor experiments, and the Mach effect experiments.

Objectives of the discussion will be to introduce a new acceleration model based on the Chameleon scalar field theory; to point out similarities to other theories, including the Alcubierre warp bubble; and to explore experimental verification.

1. J. Koury & A. Weltman, Phys. Rev. D, v69, 44026, 2004.
2. J. Koury & A. Weltman, Phys. Rev. Lett., v93, 171104-1, 2004.
3. T. Waterhouse, arXiv:astro-ph/0611816
4. https://en.wikipedia.org/wiki/Chameleon_particle

Estes Park Advanced Propulsion Workshop	Block 1	22 September 2016

Experimental Applications of Chameleon Cosmology

Glen A. Robertson
265 Ita Ann,
Madison, AL 35757

Space propulsion science has been progressing slowing over the last half century, partly due to the inconsistency between Einstein physics and quantum physics; and more so, to the incapability of propulsion engineers to understand both! The Modified Chameleon Model was developed in an attempt to bring non-classical propulsion concepts into better focus for engineers.

1. Introduction

Since indulging in the aspects of space propulsion science, I have had a hard time seeing how Einstein Relativity (ER) and its sub-theories could ever lead to the engineering of a real space drive. From my prospective, ER can only tell one that space drives are possible within the understanding of the Universe (i.e., things on a big scale) with little focus back to us mundane humans (i.e., things on a smaller scale). In recent years, there has been a focus to bring the large scale of Newtonian gravity (where today's propulsion lives) down to a smaller scale called quantum gravity (where future propulsion theories are arising). However, little (if any) engineering progress has been made.

In his paper, *"Copenhagen vs Everett, Teleportation, and ER=EPR,"* Susskind [1] provides a means to connect Quantum mechanics to Einstein Relativity and its sub-theories through a kind of non-locality called Einstein-Podolsky-Rosen (EPR) entanglement.

> *"EPR does not violate causality, but it is, nevertheless, a form of non-locality. It is most clearly seen if one imagines trying to simulate quantum mechanics on a system of classical computers. By assuming the computers are distributed throughout space and represent local degrees of freedom. The whole conglomeration is required to behave as if there were quantum systems inside the computers; systems that local observers can "observe" by pushing buttons and reading outputs. The computers will of course have to interact with each other, as they*

also would if we were simulating classical physics. But simulating classical physics only requires the computers to interact with their local neighbors." Susskind [1].

Although Susskind's idea of entanglement represents a neat way to bridge Newtonian gravity to quantum gravity through ER, when dealing with space drive theories, it may be better to indulge in the concept of coupled entanglement.

Here coupled entanglement is defined as a measure of the degree in which a system is entangled to the neighboring environment. The measure is defined as a coupling factor, where full coupling is equal to one.

For example, space drive concepts dealing with the quantum theory known as zero point energy (ZPE) may better be presented with the ZPE propulsion theory including coupled entanglement by adding a coupling factor into the ZPE propulsion equations. That is, the coupling factor represents the amount of coupling a ZPE propulsion system has to the enormous amount of ZPE energy in the Universe by only considering the coupled entanglement amount in the local or neighboring environment about the ZPE propulsion system. In other words, ZPE propulsion theories and other space drive theories have not consider entanglement to the local environment, whereby they represent a coupling factor = 1 (i.e., entanglement to the entire Universe) when it may be much smaller (i.e., coupled entanglement to the local neighboring environment). Of course understanding the correct coupling factor for any space drive may not be mathematically attainable at this time, i.e., such insights may only be attainable through experimentation. From this, it is then easy to see that coupled entanglement could solve the disagreements between ER and quantum theory by allowing coupling factors related to the degree of local entanglement to solve mathematical discrepancies. For example, although the Universe contains an enormous amount of ZPE per cubic meter, the local amount of ZPE depends on the coupled entanglement to the local environment, where the coupled entanglement is a quantum mechanism. Now the hard part, "How can propulsion engineers use the concept of coupled entanglement to investigate new propulsion theories?" Although I suspect that many new theories will emerge over time, there is one today that represents a starting point. This theory is still in its infancy and needs much work, called the "Modified Chameleon Model ", [2, 3, 4] as it is an acceleration model derived from "Chameleon Cosmology " [5].

2. Entanglement: Chameleon Cosmology and the Modified Chameleon Model

Chameleon Cosmology represents a small subtractive change to gravity related to the local or neighboring density environment, developed from a general Lagrangian where each matter field couples to a metric related to an Einstein-frame metric by a conformal transformation, where coupling factors are introduced as dimensionless constants. Unfortunately, in the papers on Chameleon Cosmology, the coupling factors are not well defined and are set to unity (=1). However, the local density environment defined in these papers provides a well-established local environment tied back to an Einstein metric, whereby the local density environment is a neighboring entanglement environment. Further, the coupling factors in Chameleon Cosmology provide a means for considering the amount of coupled entanglement to the local density environment. That is, although the Universe is large, the local density environment change across the Universe provides a local entanglement environment with coupled entanglement, where the value of a coupling factor is dependent on where you are in the Universe.

> *The universe is filed with subsystems, any one of which can play the role of observer. There is no place in the laws of quantum mechanics for wave function collapse; the only thing that happens is that the overall wave function evolves unitarily and becomes more and more entangled. The universe is an immensely complicated network of entangled subsystems, and only in some approximation can we single out a particular subsystem as THE OBSERVER.* Susskind [1].

In Chameleon Cosmology, the coupling across densities are mediated by a thin-shell mechanism (i.e., THE OBSERVER) that can be envisioned to exist about all objects to include any region defined in the Universe, where the region can include any of the states of matter, empty vacuum, or a combination of them. This observer (thin-shell) is in effect the quantum matter of the Universe, that is definable in a thin-shell medium about a defined density, where it is in entanglement to the densities on either side (internal and external to the defined density) and reacts to the changes in these densities by increasing or decreasing its thickness. The thin-shell thickness in Chameleon Cosmology effectively acts like the bridge between "folded space" in order to draw spatially distant points (densities) close to one another. The thin shell thickness in Chameleon Cosmology is basically the short black line in Susskind's [1] figure 2 and the densities are the two red dots.

> *The two red dots are maximally entangled particles and I indicate their entanglement by linking them by a short black line. The black link has some structure; for example, it distinguishes between the various maximally entangled Bell states. Despite appearances the nonlocal features of entanglement cannot be used to transmit messages super-luminally (faster than light).* Susskind [1].

Note: The Bell states are a concept in quantum information science and represent the most simple examples of entanglement. An EPR pair is a pair of qubits (or quantum bits) which are in a Bell state together, that is, entangled with each other. Unlike classical phenomena such as the nuclear, electromagnetic, and gravitational fields, entanglement is invariant under distance of separation and is not subject to relativistic limitations such as the speed of light (though the no-communication theorem prevents this behavior being used to transmit information faster than light, which would violate causality). (from Wikipedia)

General Relativity also has its non-local features. In particular there are solutions to Einstein's equations in which a pair of arbitrarily distant black holes are connected by a wormhole or Einstein-Rosen bridge (ERB). The thin-shell thickness in the Modified Chameleon Model (MCM) effectively acts like a wormhole or Einstein-Rosen bridge (ERB) (see [6]) connecting density fields. The thin-shell thickness is basically the ERB in Susskind's [1] figure 3 and the "folded spaces" are density fields. Inside the ERB resides an object constituting the location of the two observer that have jumped in, to meet.

> *At first sight it would seem that ERBs can be used to superluminally transmit signals. But this is not so; the wormhole solutions of general relativity are "non-traversible." (Non-traversibility means that two observers just outside the black holes cannot communicate through the ERB. Non-traversibility does allow them to jump in and meet in the ERB.) The similarity between figures 2 and 3 is quite intentional. The punchline of the ER=EPR joke is that in some sense the phenomena of Einstein-Rosen bridges and Einstein-Podolsky- Rosen entanglement are really the same: ER=EPR. Susskind [1].*

Susskind further states:

> *This is a remarkable claim whose impact has yet to be appreciated. There are two views of what it means, one modest and one more ambitious. The ambitious view is that some future conception of quantum geometry will even allow us to think of two entangled spins (a Bell pair) as being connected by a Planckian wormhole. The modest view first of all says that black holes connected by ERBs are entangled and also the converse; entangled black holes are connected by ERBs. But there is more to it than that. The idea can be stated in terms of entanglement being a "fungible resource." Entanglement is a resource because it is useful for carrying out certain communication tasks such as teleportation. It is fungible because like energy, which comes in different forms (electrical, mechanical, chemical, etc.), entanglement also comes in many forms which can be transformed into one another. Energy is*

conserved but entanglement is not, except under special circumstances. If two systems are distantly separated so that they can't interact, then the entanglement between them is conserved under independent local unitary transformations. Thus if Alice and Bob, who are far from one another, are each in control of two halves of an entangled system, the unitary manipulations they do on their own shares cannot change the entanglement entropy. If Alice's system interacts with a nearby environment (a first density field), the entanglement with Bob's system can be transferred to the environment (a second density field), but as long as the environment stays on Alice's side (does not cross the thin-shell) and does not interact with Bob's system (acts only with the thin-shell) the entanglement will be conserved. Susskind [1].

That is, the thin-shell in Chameleon Cosmology (CC) and the Modified Chameleon Model (MCM) act as a mediator (i.e., ERB) to conserve energy and as a mediator (i.e., EPR) to conserve entanglement between density fields. This basically entails that all objects whether stationary (CC) or moving (MCM) reside within the fundamental constitutes (i.e., exotic energy = the thin-shell energy) of a "wormhole", as defined by the wormhole solutions of general relativity. Given this, as an object moves, the "wormhole" develops around it, never having to cross an event horizon as these would form in the aft and forward wakes of the thin-shell (see ref. [6]).

3. Chameleon Cosmology and The Modified Chameleon Model

The thickness of the thin-shell about an object m of density ρ_m and radius R_m was derived in more engineering (i.e., simple) terms during the development of the Modified Chameleon Model (MCM) as

$$\Delta R_m \approx \frac{\kappa_0}{\rho_m R_m}, \qquad (1)$$

where

$$\kappa_0 \approx \frac{1}{3} M_E^2 \left(\frac{2 M_{PL}^4}{\rho_0} \right)^{1/3} \qquad (2)$$

where ρ_0 is the external density (i.e. atmospheric density about the object m) and the parameters:
Where the Reduced Plank mass (m_p) is given by

$$M_{PL} = \frac{m_p}{\sqrt{8\pi}} \approx 4.34 \times 10^{-9} \text{Kg} ; \qquad (3)$$

and Energy Scale Constant is

$$M_g \approx \left(\frac{\Lambda}{8\pi \ell_p^2}\right)^{1/4} \approx 11378 \text{m}^{-1} \tag{4}$$

where Λ is the cosmological constant and ℓ_p is the Plank length. Then it follows from the MCM derivations of Chameleon Cosmology that the subtraction to the gravitational force or Chameleon Field force is given by,

$$-F_\phi = -6\,\beta_m \left(\frac{\Delta R_m}{R_m}\right) F_\text{N}\;, \tag{5}$$

where β_m is the object's internal density coupling factor, i.e., the internal density coupled entanglement moderator of the thin-shell, and $F_N = m g_N$ is the local Newtonian gravitational force of gravity with g_N being the gravity acceleration. Under Chameleon Cosmology with MCM derivations, the gravitational force on a small object near a larger object is given by

$$F_\text{gravity} = F_\text{N} - F_\phi = \left[1 - 6\beta_N \left(\frac{\Delta R_m}{R_m}\right)\right] F_\text{N}\;, \tag{6}$$

where in Eq.(6), the subscript N is used to denote that the larger object is also the primary Newtonian gravitation force producer. And since it can be easy seen from equations (1-4) that the thin-shell thickness for the earth would be very small, ($\sim 10^{-14}$m) and $\beta_N \approx 1$ under Chameleon Cosmology, whereby the Chameleon Field force $F_g \ll 1$, so that current Newtonian gravity is not violated.

It should be appreciated that the thin-shell thickness is a quantum mechanism defined by the Planck scale parameters m_p and ℓ_p , normalized to ER (the Universe) by the cosmological constant Λ , to the external local entanglement environment of an object by the local atmospheric density ρ_0 , and to the local internal entanglement environment of the object by the object's density ρ_m. Whereby, an object's thin-shell provides the Universe entanglement (quantum) mechanism for simulating classical physics interactions between the local entanglement environments (atmosphere and object or space and space drive system). Noting that this is similar to simulating classical physics between entangled computers which only requires the computers to interact with their local neighbors.

3.1 Coupling Factors

During the development of the Modified Chameleon Model (MCM) the problem in defining the coupling factor was resolved by noting that for the earth (denoted by the subscript $\oplus$)

$$\Delta R_\oplus \approx \sqrt{\ell_p R_\oplus} \Rightarrow \ell_p \approx \frac{\Delta R_\oplus^2}{R_\oplus} \ , \tag{7}$$

but for other objects in our solar system this is not true. To correct this, equation (7) was rewritten as

$$\Delta R_\oplus \approx \beta_{E_\oplus}^2 \sqrt{\ell_p R_\oplus} \Rightarrow \ell_p \approx \beta_{E_\oplus}^4 \frac{\Delta R_\oplus^2}{R_\oplus} \ , \tag{8}$$

where for $\beta_{E_\oplus} \approx 1$, or for any object

$$\Delta R_m \approx \beta_{E_m}^2 \sqrt{\ell_p R_m} \Rightarrow \ell_p \approx \beta_{E_m}^4 \frac{\Delta R_m^2}{R_m} \ , \tag{9}$$

where β_{E_m} is the object's external density coupling factor, i.e., the external density coupled entanglement moderator of the thin-shell. Combing Eq.(9) back with Eq.(1) then gives,

$$\beta_{E_m}^2 \approx \frac{\kappa_0}{\rho_m R_m \sqrt{\ell_p R_m}} \ , \tag{10}$$

where we let all coupling factors of objects i have the same form

$$\beta_i \equiv \left(\frac{\kappa_0}{\rho_i R_i \sqrt{\ell_p R_i}} \right)^{1/2} \ , \tag{11}$$

as it can be shown that $\beta_\oplus \approx \beta_{E_\oplus} \approx 1$.
It should be appreciated that the external and the internal coupling factors are quantum factors in the same way that the thin-shell thickness is a quantum mechanism.

Now combining Eqs. (9) and (5) to the Chameleon Field force is given in terms of the coupling factors as

$$F_\phi = - \left(6 \, \beta_m \beta_{E_m}^2 \sqrt{\frac{\ell_p}{R_m}} \right) F_N \tag{12}$$

and since both coupling factors approach 1 for any object, it can be easy seen from equations (12) that the Chameleon Field force is dominated by the Planck length ($\ell_p \sim 10^{-35}$m), such that, Chameleon Cosmology does not violate Newtonian gravity theory for any gravitational object. That is, the subtraction of the Chameleon Field force does not change our view of Newtonian gravity.

4. The Modified Chameleon Model: An Acceleration Model

Chameleon Cosmology looks at the Chameleon Field force from a gravitation source on a nearby smaller objects in an external density (i.e., atmosphere) environment, where all densities are considered static. In the Modified Chameleon Model (MCM) all objects have a scalar density field, where an object's density field is only equal to its actual density when the object is static, as covered by Chameleon Cosmology. That is, the density field of an object is allowed to be timevarying. Further, MCM treats the density of the object generating the Newtonian gravitation source as a secondary external density field to the object's internal density. These density fields are taken to have same local entanglement to the thin-shell as the density environment in Chameleon Cosmology. By allowing all density fields to change, produces changes to the thin-shell thickness about the object though coupled entanglement to these density fields. Whereby, MCM investigates the Chameleon Field "acceleration" force on an object due to changes in the thin-shell thickness about an object, due to internal and external changes to local density fields. The changing density fields in MCM forces the form of the coupling factors per Eq.(11) to change as

$$\delta\beta_i \equiv \left(\frac{\kappa_0}{\beta_a \partial \rho_i \bar{R}_i \sqrt{\ell_p \bar{R}_i}}\right)^{1/2}, \qquad (13)$$

where β_a is a motion coupling factor due to acceleration, $\partial \rho_i$ is the changing density field of an object i, and R_i is the changed radius called the radial factor. These will be discussed later.

4.1 The Time Varying Density Field Model

By treating the scalar density fields like scalar potential fields, a time varying density fields can be investigated through the concept of time dilation and retardation, which is shown to eliminate the need for coupling factors by using the acceleration mediators, i.e., the mediators that cause of the coupling factors to not equal 1.

Time dilation and retardation is taken into consideration by electrical engineers when there are interfering sinusoidal electric and magnetic field; producing a small retardation of the electron motion. Retardation is a relativity slow phonon mediated process that occurs when electric and magnetic fields are sinusoidal (time-varying) and overlap in a material by an average separation distance s as described by the Lienard-Wiechert potentials (i.e., scalar potential fields [7]). This induces a small reaction time or retardation time $\Delta t = s/c$ between earlier field interactions and corresponds to a phase shift $\omega \Delta t$ and infers a retardation time $t' = t - \Delta t$, which results in unidirectional forces "on the material" in the overlapping fields.

Note: Another way of looking at the time dilation and retardation (TDR) of the thin-shell thickness is by considering the thin-shell of be composed of subatomic particles. In particle physics, many subatomic particles exist for only a fixed fraction of a second in a lab relatively at rest, but some that travel close to the speed of light can be measured to travel farther and survive much longer than expected (a muon is one example). According to the special theory of relativity, in the high-speed particle's frame of reference, it exists, on the average, for a standard amount of time known as its mean lifetime, and the distance it travels in that time is zero, because its velocity is zero. Relative to a frame of reference at rest, time seems to slow down for the particle. Relative to the high-speed particle, distances seem to shorten. Einstein showed how both temporal and spatial dimensions can be altered (or "warped") by high speed motion. (from Wikipedia) .

That is, the thin-shell thickness on one side moves faster in time than the other (time retarded) side.

4.2 Time Dilation and Retardation Model

The following time dilation and retardation (TDR) derivation is shown in more detail in ref. [3]. Under the MCM, we let the cause of the TDR comes from the motion of particulate matter (i.e., the acceleration mediators) in an object due to an applied energy potential that causes the particulate matter to move at the speed of light (even when the parent mass is moving slower) and specifically only to a small group much less than the total matter in the object and such matter that can be easily modulated without distortion to the peripheral boundary of the object or vibration of the object. That is, no visible distortion or vibration of the object would be detected. Any distortion or vibration to the object would invoke a classical energy loss in the form of mechanical, thermal or etc. energy. Although, thermal heat loss would be expected if the subatomic particulate matter is an electron, proton or neutron. Therefore, the suspect subatomic particulate matter here is probably at the quantum scale, as at this scale, matter and energy behave very differently from what much of everyday experience would lead us to expect. Under this criteria, an object's density field $\partial \rho_m$ would be changing and is given in like to Eq.(65) in ref. [3] as

$$\partial \rho_m \approx \rho_m + |\frac{\vec{F}_{\phi_a}}{\vec{F}_N}|\rho_i = \frac{3m}{4\pi \bar{R}_m^3} \quad , \tag{14}$$

where ρ_i is the density of the particulate matter and $\bar{R}_m$ is the radial factor, i.e., the change to the object's density field radius, given from Eq.(14) as

$$\bar{R}_m = R_m \left(1 + |\frac{\vec{F}_{\phi_a}}{\vec{F}_N}| \frac{m_i}{m}\right)^{-1/3} = R_m \left(1 + \left[\frac{a_i}{g_N}\right] \frac{m_i}{m}\right)^{-1/3} , \quad (15)$$

where m is the mass of the object and m_i is the total mass of the accelerated particulate matter in the object with acceleration a_i.

4.3 Time Dilation

The acceleration field force equation is then given as

$$\vec{F}_{\phi_a} \approx 6 \, \partial\beta_m \left(\frac{\partial\Delta R_m}{\bar{R}_m}\right) \vec{F}_N , \quad (16)$$

where $\partial\Delta R_m$ is the change in the thin-shell thickness that gives rise to the acceleration field force.

Now letting the ratio $\partial\Delta R_m/\bar{R}_m$ be a function of the number N of perturbations in the distribution of the particulate matter in the object over an effective time t with each perturbation occurring over the object's relaxation time $\tau \approx t/N$ corrected by a time dilation $\tau + \Delta\tau$ corresponding to a volume expansion $V_r + \Delta V_r$, which results in a dimensional translation that gives rise to the change $\partial\Delta R_m$ in the direction of any resulting motion. Such that, the acceleration field force can be given in terms of time dilation by

$$\vec{F}_{\phi_a} \approx 6 \, \partial\beta_m \left(\frac{\tau}{\tau + \Delta\tau}\right) \vec{F}_N = 6 \, \partial\beta_m \left(\frac{t}{t + \Delta t}\right) \vec{F}_N , \quad (17)$$

where the retardation time Δt reflects an interaction with an earlier event from the current time t and corresponds to a phase shift

$$\Delta\omega_m \approx \omega\Delta t . \quad (18)$$

4.4 Retardation

Retardation implies that there is a retarded or past density change $(\partial\rho_m)_R$ and a non- retarded or current density change $(\partial\rho_m)_{NR}$, which gives rise to a motion coupling factor between the object and the Newtonian gravitation object to not be identical. This difference implies that there exists both a retarded internal coupling factor $(\beta_m)_R$ and a retarded Newtonian internal coupling factor $(\beta_N)_R$, given by

$$(\beta_m)_R \approx \partial\beta_m \sin(\omega t + \phi - \Delta\phi_m) ; \quad (19)$$
$$(\beta_N)_R \approx \partial\beta_N \sin(\omega t - \Delta\phi_m) , \quad (20)$$

noindent where ωt is the phase between events and ϕ is the phase between the changing coupling factor $\partial\beta_m$ and the changing Newtonian mass to field coupling factor $\partial \beta_N$ relating to the non-retarded or current density change $\partial \rho_m$ of the object, given by

$$(\beta_m)_{NR} \approx \partial \beta_m \sin(\omega t + \phi) \; ; \quad (21)$$
$$(\beta_N)_{NR} \approx \partial \beta_N \sin(\omega t) \; , \quad (22)$$

where the subscript ($_R$) implies retarded and the subscript ($_{NR}$) implies non-retarded. Now let

$$\theta_m \approx 6 \, \partial \beta_m \left(\frac{t}{t + \Delta t} \right) \; , \quad (23)$$

defined as the local fifth force coefficient, where equations (20-23) provide a phasing of the local fifth force coefficients as

$$\theta_m \approx 6 \, \partial \beta_m \sin(\phi) \; . \quad (24)$$

Then noting that,

$$\sin(\phi) = \frac{t}{t + \Delta t} = \frac{\partial \Delta R_m}{\bar{R}_m} \ll 1 \quad (25)$$

and that when the $\sin(x) \ll 1$, $\sin(x) \approx x$, such that the phase

$$\phi \approx \frac{t}{t + \Delta t} \; . \quad (26)$$

Then using Eq.(17), the acceleration field force can be given in terms of a phase factor as

$$\vec{F}_{\phi_a} \approx 6 \, \partial \beta_m \phi \vec{F}_N \; , \quad (27)$$

where from equations (27) and (13), the internal coupling factor

$$\partial \beta_m \approx \frac{1}{6\phi} |\frac{\vec{F}_{\phi_a}}{\vec{F}_N}| = \frac{1}{6\phi} \left(\frac{a_i}{g_N} \right) \approx \left(\frac{\kappa_0}{\beta_a \partial \rho_m \bar{R}_m \sqrt{\ell_p \bar{R}_m}} \right)^{1/2} . \quad (28)$$

Whereby, the Eq.(28) gives the thin-shell change

$$\partial \Delta R_m \approx \phi \bar{R}_m \approx \frac{1}{6\phi} \left(\frac{a_i}{g_N} \right) \left(\frac{\kappa_0}{\beta_a \partial \rho_m \bar{R}_m \sqrt{\ell_p \bar{R}_m}} \right)^{-1/2} \bar{R}_m \; , \quad (29)$$

by noting Eqs. (25) and (26) to give the phase factor

$$\phi \approx \frac{1}{6\phi} \left(\frac{a_i}{g_N} \right) \left(\frac{\kappa_0}{\beta_a \partial \rho_m \bar{R}_m \sqrt{\ell_p \bar{R}_m}} \right)^{-1/2} = \frac{1}{6 \, \partial \beta_m} \left(\frac{a_i}{g_N} \right) \; . \quad (30)$$

The main point of Eq. (30) is that the phase factor inherently carries the motion factor β_a and the internal coupling factor $\partial \beta_m$. That is, once knowing the phase factor, the motion factor and the internal coupling factor can be derived. However as is shown in the following knowing the motion factor and the internal coupling factor is not required, once the phase factor is known.

4.5 MCM Rocket Model

A rocket is a two density field model, i.e., the changing propellant causes the rocket to have a changing density field and the gas flow through the nozzle induces a new changing density field, such that a two density field approach is required [3, 4].

For a rocket, the thrust can be given by

$$T \approx (m_i - m_{ex})a_r , \qquad (31)$$

where m_i is the initial mass of the rocket and m_{ex} is the exhausted mass.

In ref. [8], it was shown that the MCM model can be simplified to the rocket acceleration as

$$a_r \approx 6 \left[\frac{\phi^3 \ell_p^{-1/2}}{\left(\bar{R}_r^{-1/2} - \bar{R}_{gas}^{-1/2} \right)} \right]^{1/2} g_N , \qquad (32)$$

Where the rocket radial factor $\bar{R}_r$ is the nozzle throat radius (the active interface between the rocket and nozzle) and the gas radial factor $\bar{R}_{gas}$ is $\sqrt{2}$ times the exhaust nozzle radius (the active interface between the nozzle and the external density field), and where the phase is given by

$$\phi \approx \left[1 + \left(\frac{v_{gas}}{\bar{R}_{gas}} \right) \left(\frac{m_{ex}}{\dot{m}} \right) \right]^{-1} . \qquad (33)$$

The active interfaces are the locations where mass is flowing out of one density field and into another.

For a rocket, the hot gas relaxation time

$$\tau \approx \frac{\bar{R}_{gas}}{v_{gas}} , \qquad (34)$$

where v_{gas} is the hot gas velocity, and the rocket's mass flow retardation time

$$\Delta \tau \approx \frac{m_{ex}}{\dot{m}} , \qquad (35)$$

where $\dot{m}$ is the propellant mass flow rate crossing the throat.

4.6 MCM Single Object Model

Extrapolating the case of a rocket to a single object with no active interfaces (i.e., no expelled mass), we let an object's acceleration

$$a_m \approx 6 \left[\frac{\phi^3 \ell_p^{-1/2}}{\left(\bar{R}_1^{-1/2} - \bar{R}_2^{-1/2} \right)} \right]^{1/2} g_N , \qquad (36)$$

where $\bar{R}_1$ is the radial factor in the direction of motion produced by an accelerated particulate matter of total mass m_i and $\bar{R}_2$ is the radial factor in the opposition direction of motion cause back the relaxation accelerated particulate matter of total mass m_i, i.e., the particulate matter in the object is accelerated in one direction and allowed to relax (i.e., the acceleration force is turned off) back across the object in the opposite direction. Noting that $\bar{R}_1$ should be smaller than $\bar{R}_2$, so that the acceleration is positive.

The phase factor then becomes

$$\phi = \phi_1 - \phi_2 \approx \left[1 + \left(\frac{v_1}{\bar{R}_1} \right) \left(\frac{m_i}{\dot{m}_1} \right) \right]^{-1} - \left[1 + \left(\frac{v_2}{\bar{R}_2} \right) \left(\frac{m_i}{\dot{m}_2} \right) \right]^{-1} . \qquad (37)$$

Or noting acceleration $a = dv/dt$

$$\phi = \phi_1 - \phi_2 \approx \left[1 + \left(\frac{a_1 dt_1}{\bar{R}_1} \right) \left(\frac{m_i}{\dot{m}_1} \right) \right]^{-1} - \left[1 + \left(\frac{a_2 dt_2}{\bar{R}_2} \right) \left(\frac{m_i}{\dot{m}_2} \right) \right]^{-1} . \qquad (38)$$

Noting that ϕ_1 should be smaller than ϕ_2, so that the acceleration is positive. Also note that when $a_1 = a_2$, $\bar{R}_1 = \bar{R}_2$ from Eq.(14), $dt_1 = dt_2$, and $\dot{m}_1 = \dot{m}_2$. Whereby the phase factor and the object's acceleration is zero.

4.6 Gravity and Phase Factor

Equation (38) can be reduced by looking at the relaxation phase factor ϕ_2 when it is a result of gravity ($a \equiv 1$, $g \equiv 2$), where the phased factor is given as

$$\phi = \phi_a - \phi_g \approx \left[1 + \left(\frac{a_m}{\bar{R}_a} \right) \left(\frac{m}{\dot{m}_a} \right) dt \right]^{-1} - \left[1 + \left(\frac{g_N}{R_m} \right) \left(\frac{m}{\dot{m}_g} \right) dt \right]^{-1} ; \qquad (39)$$

note that $\bar{R}_2 = R_m$ from Eq.(14) as $\partial \rho_m = \rho_m$, and the time dt for the two phases are the same as the two accelerations a_m and g_N are acting on the object at the same time.

Now at some time $dt = t_h$, $\phi_g = \phi_a$ and the object falls back to earth with a phase factor

$$-\phi_g \approx -\left[1 + g_N \left(\frac{t_h}{R_m}\right)\left(\frac{m}{\dot{m}_g}\right)dt\right]^{-1} \quad (40)$$

Now using Eq.(36) with the object's acceleration $a_m = g_N$ and with $\bar{R}_1 = \bar{R}_a = 0$ as the acceleration stops, where

$$g_N \approx 6\left(-\phi_g^3 \sqrt{\frac{R_m}{\ell_p}}\right)^{1/2} \quad g_N \Rightarrow 6\left(-\phi_g^3 \sqrt{\frac{R_m}{\ell_p}}\right)^{1/2} \approx 1. \quad (41)$$

or

$$-\phi_g \approx \left(\frac{1}{36}\sqrt{\frac{\ell_p}{R_m}}\right)^{1/3}. \quad (42)$$

Combining Eqs. (40) and (42) yields the acceleration of gravity at time t_h as

$$g_N \approx \left(\frac{R_m}{t_h}\right)\left(\frac{\dot{m}_g}{m}\right), \quad (43)$$

which when combined back with Eq.(41) yields

$$\phi_g \approx \frac{1}{2}, \quad (44)$$

which when combined with Eq.(40) (ignoring the directional - sign) yields

$$\dot{m}_g \approx m\left(\frac{t_h}{R_m}\right) g_N. \quad (45)$$

Now noting that $\dot{m}_g = m/t_h$, Eq.(45) (ignoring the directional - sign) yields

$$t_h \approx \sqrt{\frac{R_m}{g_N}}, \quad (46)$$

where when combined back with Eq.(45) (ignoring the directional - sign) yields

$$\dot{m}_g \approx m\sqrt{\frac{g_N}{R_m}}. \quad (47)$$

As a check, from similarity, at time $dt = t_h$ with $\phi_g = \phi_a$,

$$\dot{m}_a \approx m\sqrt{\frac{a_m}{\bar{R}_a}} = \frac{m}{t_h} \Rightarrow t_h \approx \sqrt{\frac{\bar{R}_a}{a_m}}, \quad (48)$$

which when combined with the acceleration phase factor of Eq. (39) yields $\phi_a \approx 1/2$.

Now combining Eqs. (47) and (48) with Eq.(39) the phase factor is reduced to

$$\phi = \phi_a - \phi_g \approx \left(1 + dt\sqrt{\frac{a_m}{\bar{R}_a}}\right)^{-1} - \left(1 + dt\sqrt{\frac{g_N}{\bar{R}_m}}\right)^{-1}, \quad (49)$$

or rewriting Eq.(49) in the form of Eq.(38), the phase factor of an object with internal matter acceleration is given by

$$\phi = \phi_1 - \phi_2 \approx \left(1 + dt_1\sqrt{\frac{a_1}{\bar{R}_1}}\right)^{-1} - \left(1 + dt_2\sqrt{\frac{a_2}{\bar{R}_2}}\right)^{-1}, \quad (50)$$

which can be combined with Eq.(34) to give the net acceleration of a non-classical (no ejected mass) thrusting system.

5. Conclusions

As mentioned in the introduction, the MCM [2-4, 6 and 8] is a work in progress, where the visual and math models are based on the thin-shell mechanism in Chameleon Cosmology [5, 6] converted to an acceleration model and further converted to a more engineering series of equations. Chameleon Cosmology has had much work given toward its application to quantum phenomena as the Casimir Force (e.g., see [9] and its references) and is a Planck scale model. Susskind?s [1] entanglement approach to link Einstein Relativity to quantum theory can easily be applied to the thin-shell mechanism as the thin-shell is in an entanglement to the density both internal and external to an object. Whereby, Chameleon Cosmology provide another link from Einstein Relativity to quantum theory for fixing ER toward investigating and understanding new propulsion models. The MCM equations in both the more science form having coupling factors and the more engineering form not including coupling factors have been applied to the same solid rocket motor problem [3, 4, and 8] with the correct acceleration derived in each case. Beyond this no other acceleration device has been analyzed. The engineering acceleration Eq.(36) and the phase factor of Eq.(50) derived in this paper will require validation by testing. Changes to the equations in general would not be unexpected at this point, but the equations presented provide a good start toward modeling and understanding new propulsion concepts that reside outside the classical model. It is noted that the MCM of propulsion, i.e., Eq.(48), requires the propulsion systems to be non-linear, i.e, the forward and reverse accelerations are non-equal. This may require metamaterials, a material engineered to have a property that is not found in nature, to be developed to fully utilized the MCM model to the fullest extent possible.

This paper takes a new look at the Modified Chameleon Model by drawing in entanglement through the work of Leonard Susskind, who lays the foundation that entanglement is the bridge between Einstein Physics and Quantum Physics. A discussion entailing the entanglement connection is presented followed by an overview of the Modified Chameleon Model. Lastly, the acceleration equation in the Modified Chameleon Model is revisited and a general form of the acceleration phase factor is developed for use with non-classical propulsion concepts that have no mass ejection.

6. Discussion

Woodward: You showed inertia as a sort of offset due to an asymmetry in the expanding universe. Yet the expansion is isotropic in the universe. How do you get such an asymmetry?

Robertson: I am not showing the matter of the universe, but the product of density and acceleration.

Woodward: But the acceleration is isotropic, too.

Robertson: If you are at the edge of the universe, looking back, you don't see anything beyond you.

Woodward: No, you just see more universe. It is all isotropic about every point.

Robertson: My assumption is that if you go beyond the edge of the universe, there is nothing.

Woodward: The edge of the universe isn't defined that way. There is always more beyond it.

Robertson: I take your point, but right or wrong, this is how I am visualizing things in this model.

Hathaway: Can you remind me where the Planck length comes into this chameleon model?

Robertson: It came from me. I inserted it based on an assumption relating a parameter to the mass of the earth.

Hathaway: Can you explain how you were getting a force on a density field, and what that means? I don't understand the concept.

Robertson: From our standpoint, the density field is the space drive.

Rodal: I would like to comment on your coupling parameters. Remember the old model of planetary motion involving epicycles? There was a lot of data to feed the model of epicycles, and every new discovery involved more epicycles. But the epicycle model was quite good. I see your coupling parameters as like the epicycles. The test of a theory is not whether it can be made to match data – it always can with an appropriate choice of parameters. The real test of a theory is whether it can make new predictions outside the known field of data, that can be verified in data. Does your theory make a prediction outside known data?

Robertson: I have not done that. I am just an engineer that assumes the model. But I leave it to physicists to determine whether the model is valid. I like this model because I think it can be used as a common model. And I just don't have the time to look in detail at these things.

Williams: You mentioned to George that you want input to improve your model. I am still having trouble with the density field. We have various pictures and concepts we use to organize our understanding. It would be good if you could tie this back to something that is understood. It's hard when you simultaneously introduce a new concept and start doing calculations with it.

Robertson: I suggest you go back to the original paper to understand the density field. I have added the assumption that the density field depends on accelerations within the density field. Somebody needs to go back to the original paper and put in my changes to alter their theory.

Williams: But I just want to know what the concept is. What is a density field? Since it is the basis of your talk, can you offer a picture?

Robertson: I don't really understand the math. I would rely on people like you to help me understand the math. I am really asking for your help. But I have this simple picture of the density depending on acceleration. Maybe density is the wrong term? But I am taking it out of the chameleon cosmology. We can call it whatever you want.

Woodward: In an earlier slide, you were showing acceleration with no mass ejection, just density fields. I am curious how you get that and treat that?

Robertson: What you are operating on is the thin shell. The acceleration comes from that.

Woodward: That sounds like making your car move by pushing on the dashboard.

Robertson: I have claimed in the past you can do that. But I really want to show your theory and others can be cast in these terms.

Woodward: My theory is already within existing physics, I just have a new effect. Why would I want to introduce density fields and all these other parameters when I can simply use existing physics to describe my theory?

Robertson: It might not be so useful for your theory, but wouldn't it be a useful common framework for evaluating lesser-known theories than yours.

Woodward: Only theories that involve your density field.

Robertson: My impression from attending these conferences for years is that all these theories are talking about the same thing. My approach allows a way to express these in a common framework.

Woodward: What you have done is taken an outside cosmological model with questionable assumptions and suggested it as a framework that all models should be pushed into.

Robertson: I'm not sure this is the right model, we just need a common model so that funders can understand what we're doing. If this is not the right model, I challenge others to come up with a better one.

Williams: You would not want to use this to capture things like the Mach effect that purport to be within general relativity. It could only be for new physics.

Robertson: This is the sort of derivation that could get into an aerospace book, but the Mach effect could not. I have not seen anything today that you could not put into the thin-shell chameleon model. The equations may change but it would fit.

Laursen: Are you familiar with Kane's approach at Stanford? He developed a framework in which any dynamics problem could be formulated. It sounds similar.

Robertson: If you fit your parameters into this model, it will help improve your experiment, because the free parameter are experimentally determined. So I am not saying put your theory in this model, put your experiment in this model.

Meholic: Have you seen Richard Obussey's PhD thesis on the Alcubierre metric? He had a picture of expanding dimensions to propel an object. This sounds similar.

Robertson: The equations may change but the model stays the same.

Williams: In terms of parameterizing theories, if you have enough free parameters you can fit anything. If you make enough assumptions to get to an answer, why not just assume the answer and save a step?

Robertson: The parameters are not from the theory but from the experiment.

Williams: That's fine but if the space of parameterization is too big... How many parameters do you have?

Robertson: I couldn't answer that, but not too many. Perhaps 4. Coupling factor to internal field, coupling factor to external field, ... phase factor, radial factor... but it should come out of experiment. If you can't put your experiment derived from your theory into this model, then it is suspect.

Woodward: I don't understand how experiments can be suspect. If an experiment is done well, it can be taken as factual. Whereas, models can be suspect.

Robertson: Your experiment fits well in this model. Your experiment changes the density field of the PZT.

Woodward: I'm not changing the density field of the PZT. I am applying a voltage to a PZT stack, producing electro-mechanical motion of known physical elements. Your density fields are not known physical elements. Nobody has ever measured a density field. If it's a calculation of an assumption it does not have the same status as an observed fact.

Robertson: My main assumption is we need a common base to give to people who don't understand theoretical physics.

Woodward: That's a worthy goal. Why not try to do it without density fields and chameleon stuff?

Robertson: Nothing else has worked.

Woodward: That's not so. There is a parameterization of gravitational theories developed by Ken Nordtvedt and Cliff Will back around 1970. It is still in common use: Parameterized Post-Newtonian Theory. You could find the appropriate terms of those equations and map them to real effects. They have already done for gravity what you would like to do, without needing to introduce any new fields.

Robertson: I am presenting a common model for your evaluation but you must decide its utility for yourself.

Broyles: Do you have a simple, step-by-step block diagram to walk people into this model? Let's say I have a black box that produces some effect. How do I parameterize what I measure with it in your model? Your equations don't help us in that regard. You would need the detailed steps to present to an organization you hope would fund this.

Robertson: I will have to think about putting something like that together.

Turner: I work at NASA and have some experience with the issues of explaining proposals. I agree with Tony's assertion there is a need to compare alternative approaches at a level that could be easily understood by a wide range of people. Whether it's a common model or impartial review by competent scientists, that would be good.

Tajmar: One point is that PPN is the valid way to depart from general relativity, which is the accepted theory. A second point is that, in terms of propulsion, a proposal should give two numbers: power to thrust ratio and specific impulse. Those are the parameters to use to compare approaches. In general, these figures of merit already exist from an engineering perspective for thrusters, and from a theory perspective for gravity. There is nothing else.

Bushnell: There are probably more theories than theorists. I have tried to work through a lot of these exotic proposals. I have talked to the best scientists at Harvard, Princeton, Michigan, whatever. And they tell me the woods are full of this stuff. There aren't enough hours in a year to look through it all, and it probably wouldn't be of value. There is something here broken between conventional physics and this field, and conventional physics faces many big questions. So how do we parse which of any to go forward with? But conventional physicists won't touch this stuff. If this meeting does nothing, we should get the theory squared away. There are lots of issues in the experiments. But with unverified assumptions, how can you possibly believe anything?

Broyles: You look for specific, repeatable, verifiable experiments. The basics.

Meholic: It's also important to have people who understand the data. In the military world, they want to know what capability is promised. There is no clear mission for this in the NASA realm, but possibly in the military realm.

Fearn: As a theorist, I had to look up the meaning of specific impulse that Martin mentioned. It is a ratio of thrust to mass ejection. But our experiments don't eject anything. Why can't we have thrust over power in?

Meholic: Yes, specific impulse is meaningless in this application. It is related to expelled matter. But in the electrical propulsion world, they use just what you say. I have always defined specific impulse as the time to burn one unit mass of propellant to produce one unit of thrust. The higher the number, the more efficient.

Cole: The electric propulsion thrust to power ratio is sometimes called alpha. But the key parameter is how much energy the power supply can produce before you "run out of gas".

Broyles: It took decades to understand the physics of flight after the Wright brothers proved it experimentally. So I am not concerned if we are seeing new physical effects without a clear theory.

Tajmar: Or superconductivity. We use it all the time and haven't got a clue how it works!

References

[1] Susskind, Leonard, *Copenhagen vs Everett, Teleportation, and ER=EPR*, arXiv:1604.02589 v2 [hepth], 23 Apr. 2016.

[2] Robertson, Glen A., *Engineering Dynamics of a Scalar Universe, Part I: Theory & Static Density Models*, Lecture Series paper Proceedings of Space, Propulsion & Energy Sciences International Forum, edited by Glen A. Robertson, AIP Conference Proceedings, CP1103, Melville, New York, (2009).

[3] Robertson, Glen A., *Engineering Dynamics of a Scalar Universe, Part II: Time-Varying Density Model & Propulsion*, Lecture Series paper in Proceedings of Space, Propulsion & Energy Sciences International Forum (SPESIF-09), edited by Glen A. Robertson, AIP Conference Proceedings, CP1103, Melville, New York, (2009).

[4] Robertson, Glen A., *The Chameleon Solid Rocket Propulsion Model*, in these proceedings of Space, Propulsion & Energy Sciences International Forum (SPESIF-10), edited by Glen A. Robertson, AIP Conference Proceedings, Melville, New York, (2009).

[5] Khoury, Justin and Weltman, Amanda, *Chameleon Cosmology*, arXiv: 0309411v2 [astro-ph], 1 Dec. 2003 and *Chameleon Fields: Awaiting Surprises for Tests of Gravity in Space* arXiv: 0309300v3 [astro-ph], 9 Sept. 2004.

[6] Robertson Glen A. and Pinheiro, Mario J., *Vortex Formation in the Wake of Dark Matter Propulsion*, Physics Procedia, **20**, Elsevier Science, 2011.

[7] Lienard-Wiechert potentials, Wikipedia, 6 Sept. 2016.

[8] Robertson, Glen A., *Propulsion Physics under the Changing Density Field Model*, JANNAF, 2012 and presented at the Advanced Space Propulsion Workshop 2012.

[9] Almasi, A., Brax, Philippe, Iannuzzi, D. and Sedmik, R. I. P., *Force sensor for chameleon and Casimir force experiments with parallel-plate configuration*, arXiv:1505.01763v1 [physics.ins-det], 7 May 2015.

	Fuel Problem		Theory	✔	Existing Physics
✔	Time-Distance Problem		Experiment		New Physics

Alternate Concept: Tri-Space Model

- Greg Meholic will lead a discussion on an alternate conceptual viewpoint to thinking about spacetime that could facilitate faster-than-light propulsion

Issue Summary:

This session will be a departure from previous sessions. Instead of discussing a particular experiment, or a particular mathematical theory, Greg will discuss an alternative viewpoint that he calls the Tri-Space Model of the universe. It is inspired by the mathematics of the expression for relativistic energy, $E = m_0 c^2 / \sqrt{1 - v^2/c^2}$, which is a real number for speeds below light, diverges when the speed equals the speed of light, and becomes imaginary for speeds above light.

Greg has applied his model to the origin of inertia. At Estes Park he will discuss how the Tri-Space model relates to the known particles and fields in physics and cosmology, and why it has implications for faster-than-light travel.

Greg would like to share this alternative way of thinking about spacetime and about faster-than-light propulsion, and also to gather your feedback and ideas.

Estes Park Advanced Propulsion Workshop	Block 3A	22 September 2016

RECORD OF THE PROCEEDINGS:

a session on

The Tri–Space Model of Spacetime and the Universe

led by Gregory V. Meholic

Greg provided a change of gears and change of perspective at the workshop by asking us to reconsider our conceptions of the propulsion problem. He has developed over time a conceptual, notional model that to tie together some underlying phenomena of nature. He brings an engineer's visual perspective to these considerations, but in a way that does not involve mathematical equations. So this is a presentation to allow him to share his visualization of some of the aspects of spacetime and the propulsion problem.

This picture was inspired when Greg stumbled upon an old report on tachyons [1] by Edward Puscher of Rand Corporation. That report contained the relativistic equation for energy of a body of rest mass m moving with speed v:

$$E = \gamma mc^2 = \frac{mc^2}{\sqrt{1 - v^2/c^2}} \qquad (1)$$

Any massive body accelerating from rest must have $v < c$, and so the denominator in the energy expression is always positive. The quantity γm is sometimes called the relativistic mass. When $v \ll c$, $E \simeq mc^2 + mv^2/2$, the non-relativistic kinetic energy plus rest mass energy.

When speeds $v > c$ are considered in (3), then the energy becomes imaginary. It is therefore hypothesized that such a particle must have an imaginary mass, so that the energy stays real. Therefore the range of energies (for positive speeds) is

$$\begin{aligned} E &= \frac{mc^2}{\sqrt{1 - v^2/c^2}} \quad , \quad v < c \\ E &= \frac{mc^2}{\sqrt{v^2/c^2 - 1}} \quad , \quad v > c \\ E &= \infty \quad , \quad v = c \end{aligned} \qquad (2)$$

Therefore the entire domain of values of energy E for any v comprises two domains with finite values of energy: above and below the speed of light.

A third domain consists of diverging energy values when approaching the speed of light from above or below. The figure below allows a negative energy region as well, in the spirit of Dirac. The negative velocity region accounts for particles moving left or right, and it is understood these speeds are projections of three-dimensional velocity vectors.

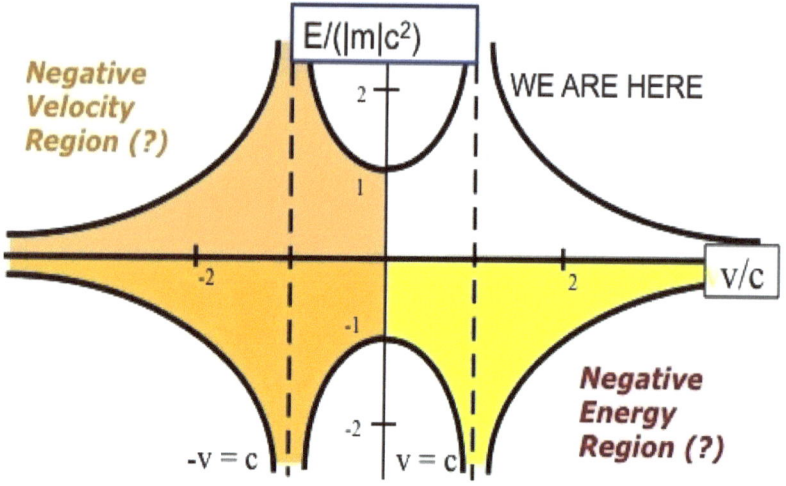

Figure 1: Basis of the Tri–Space Universe.

It is hypothesized that the laws of physics may be different in the 3 regions; 3 overlapping realms in the same space. We only have access to the $v < c$ region. Particles inhabiting the $v > c$ region are tachyons, although none have ever been observed.

The $v = c$ realm is inhabited by electromagnetic waves and gravity waves. The $v > c$ realm has the curious property that higher speeds correspond to lower energies. Energy would be required to slow down a tachyon to the speed of light.

The 3 realms are envisaged as mirroring their representation in velocity space: the subluminal and superluminal regions adjoin at a "luminal" sheet where $v = c$. The sheet can be deformed similar to how we picture deformation of spacetime under gravity. It is suggested that the sheet could be identified with quantum foam or the quantum vacuum.

> **Williams**: Don't we also have quantum foam in the subluminal realm? What is the connection to moving at the speed of light?
>
> **Meholic**: It should become clear later.
>
> **Williams**: And aren't the 3 realms sharing the same space? I can have a particle moving submluminal and a hypothetical

superluminal particle, and couldn't they meet in the same space?

Fearn: I think time stops for the luminal particles, and goes backward for the superluminal ones.

Meholic: Let me put that discussion off and we will come back to it.

Greg introduces a model of the electron developed by Richard Gautier. It hypothesizes that the electron is a manifestation of a sub-particle that executes spatial motion on a lengthscale similar to the size of the electron. This model is an alternative explanation for the zitterbewegung phenomena *[Ed: that Schroedinger showed to be implied by the Dirac equation: a relativistic interaction between an electron's translational motion and spin should lead to a violent oscillation of the particle at very high frequencies and over distances of roughly one Compton wavelength. So far, this has not yet been observed.]*

Williams: Are you treating the electron as a particle? As opposed to a wave?

Meholic: Yes, that is his model.

It is conjectured that perhaps the fundamental sub-particle that constitutes the electron in the Gauthier picture, called by Gauthier a "TEQ", could also be a constitutent of other types of matter, and Gauthier has made some investigations in this area. Since the Gauthier model involves the sub-particle moving alternatively faster and slower than light, Greg suspects this conjectured particle could provide a link to the superluminal realm. Perhaps all mass has a superluminal component in an analogous way.

Williams: What are the parameters of the TEQ that produce the other parameters of the electron like charge and mass?

Mathes: The parameters are in Gauthier's paper. We can send it to you. But there are about 6 parameters. But this particular model does not have charge in it yet.

Jansson: So you are saying this TEQ goes across the luminal boundary?

Meholic: Yes

Bushnell: The wavefunctions for tachyons are sub-luminal, so the wave function could provide the coupling between the superluminal and subluminal realms; through a sort of quantum entanglement.

Greg goes on to suggest the luminal boundary has a finite "thickness" and relates this to space and time in a sketch at the whiteboard. He emphasizes he is just brainstorming with this picture and does not present these as in any way complete results.

Fearn: This sounds very similar to Cerenkov radiation.

The history of fluid conceptions of spacetime, such as the aether, are reviewed. It is suggested that a fluidic picture may have relevance for today's physics. Moreover, the Tri-Space model invites some quasi-mechanical conceptions, "physical analog", of the phenomena underlying charge and mass.

Williams: The references shown on the slide for papers on fluidic spacetime are not really mainstream journals. So we are unsure if there is a renaissance of the aether, for example.

Meholic: These may be some dated references.

Rodal: What are the properties of the spacetime fluid?

Meholic: We are coming to that. Stay tuned.

Rodal: Also, you refer to the fluid properties of spacetime, and also to the aether, but the aether is a solid. Are you not distiguishing a solid and a fluid? The speed of light was related to its putative modulus of elasticity. The aether has a very high modulus of elasticity. There are mathematical distinctions between a solid and a fluid.

Meholic: The answer to your question is that it's a fluid in this model, because of the Gauthier assumptions.

Tajmar: Earlier you said the aether approach could resolve some questions that quantum mechanics cannot. What are those?

Meholic: I must admit I was just quoting that unverified and I cannot name any.

Greg goes on to describe fluidic interpretation to masses and fields, and a similarity is noticed to the quantum vacuum. Density, compressibility, viscosity, etc, are ascribed to spacetime.

Williams: When you talk about spacetime, do you mean the subluminal, luminal, what?

Meholic: The luminal.

Williams: So when you talk about compressibility, you aren't referring to subluminal or superluminal states? Aren't we in subluminal realm?

Meholic: We are experiencing subluminal speeds in luminal spacetime...Because we are receiving light from the lights...It's a convoluted way to think about this.

Williams: Which of the 3 realms corresponds to spacetime?

Meholic: All of them. It's the same space, but with 3 possible velocities at once. Depending on your energy state in that continuum, you would experience subluminal, luminal, or superluminal.

Bushnell: I question the validity of your continuum model. It implies a certain mean free path. The fundamental reality must be quantum.

Meholic: Point taken, I am just trying to keep things at the simplest meaningful level.

Fearn: I think Dennis is saying, for example, you can't define an index of refraction for two atoms. There is a limit to the continuum analogy. Is it an appropriate conception?

Meholic: There is no rarification. TEQs are everywhere and they are right next to each other.

Fearn: Perhaps you have a tight coupling due to short timescales?

Tajmar: You are saying the TEQ density is uniform everywhere?

Greg departs into some conjectures regarding the superluminal space. It is emphasized that the rest masses are presumed imaginary to keep the energy real.

Williams: I just want to point out that the imaginary mass is an artifact of extending the subluminal energy relation to superluminal speeds. The theory of relativity does not actually dictate whether the metric signature is $+---$ or $-+++$. If you know you are going to allow superluminal speed, you can adopt a convenient normalization for the energy relation that keeps the energy positive. The point is I would not read too much into the notion of imaginary mass.

Greg then describes a picture of interaction at the boundary between the subluminal and superluminal spaces.

Williams: You describe this as a spatial boundary, but didn't you say the 3 realms share the same physical space? And you describe as a plane something moving at the speed of light. So I am having trouble with the concepts.

Meholic: I am not surprised, it took me 27 years to come up with this!

Greg goes on to describe an interpretation of gravity in terms of the membrane between the realms. He conjectures that the 3 realms could be tied to different phase states of the different states of motion. Could a phase change lead to superluminal velocities? Greg sees many phenomena of nature "explainable" in terms of the tri space model.

– discussion of changing between the three states of motion –

Greg suggests instead of thinking of acceleration and deceleration, to instead think of changing states of motion as if entering a different space.

Fearn: Perhaps you could think about it like changing the refractive index of space.

Greg showed some of his intuitive visualizations of gravity in these terms. He conjectures that spacetime provides a viscous resistance to the motion of bodies, that accounts for inertia. It is treated as something continuous on a subatomic scale, even. In Greg's vision, electric charge is related to the TEQs, and they are imagined to be corkscrewing through different realms. He visualizes that electromagnetic waves are also TEQs.

Greg concludes by thanking us for indulging his intuitive pictures, challenging us to keep open minds and defy convention to make progress.

Reported by L. Williams

Reference

[1]. Puscher, E. A., "Faster-than-Light Particles: A Review of Tachyon Characteristics, Paper N-1530-AF, Rand Corporation, Santa Monica, CA, 1980. http://www.rand.org/content/dam/rand/pubs/notes/2009/N1530.pdf

Observations & Summing Up

L.L. Williams & H. Fearn

This was the first meeting of its kind devoted to the propulsion problem, and we are gratified that the unique format and approach proved to be a success as judged by the participants. Our focus was on the process of reviewing and assessing new ideas, and we threw a wide net to feed that process. We feel we provided the time and space necessary for promising new ideas to get a hearing, win adherents, and effect progress.

These Proceedings present papers on a range of topics. However, there was a focus at the meeting on propellantless propulsion experiments. There were several sessions devoted to this topic. A keen interest in this topic was shared by many participants, and the schedule was altered at the last minute to accommodate an extended group discussion on it. Other papers in the Proceedings stand on their own. Here, we draw together the combined implications of the Estes Park propellantless propulsion results. Perhaps the most significant outcome of the workshop was a tentative concordance among different research groups that a non-zero thrust signal may be present in one or the other of the two propellantless thrust device designs considered in Estes Park.

One design is the Mach effect device, which relies on an understanding of inertia as being due to the cumulative gravitational interaction of all the mass-energy in the universe. A putative propellantless thrust is produced with a clever electro-mechanical manipulation of the device. The Mach effect theory and first experiments were developed by Woodward. A host of other researchers have tested similar devices or other devices built by Woodward. That includes Hathaway, Tajmar, and Buldrini, who present detailed results in this volume.

Buldrini finds that the Mach effect devices exhibit a characteristic profile of thrust versus time. It is shown in Figure 1, taken from Buldrini's paper in this volume.

The second design is an asymmetrically-shaped resonant RF cavity that is reported to produce thrust when microwaves are emitted inside. This is commonly called an EM-drive, for electromagnetic drive. March, who reports results in this volume, calls it a Q-thruster, for quantum-vacuum thruster. Here, March assumes the thrust produced by the device is due to an interaction with the quantum vacuum. This hypothesis has been questioned by other researchers, and all parties agree that it is impossible to produce such thrust from electromagnetic effects alone, no matter the shape of the cavity. The device was first proposed by Shawyer, and his theoretical understanding of electromagnetism is suspect. However, experimentalists seem to be seeing a repeatable effect. Woodward suspects that EM-drive devices are operating on a Mach effect principle, and Montillet attempts to provide such an explanation in this volume.

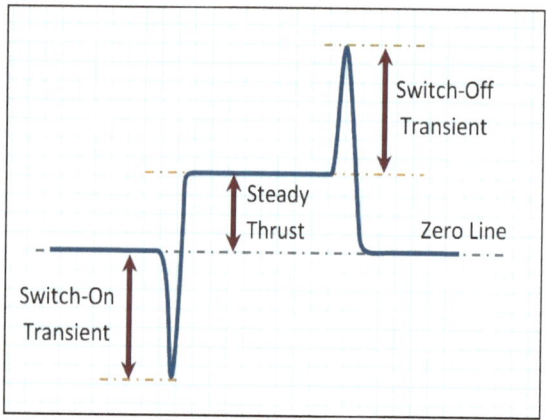

Figure 1: Schematic diagram of the typical Mach-effect thrust signal.

The typical thrust levels of the Mach effect devices reported by Estes Park participants were 0.1 to 1 μN. Thrust levels reported for the EM-drive were 10-100 μN. These results should be normalized to the power required for a photon rocket, which can set a lower limit from the thrust achievable by electric power.

The maximum photon momentum output per power input is just the ratio of momentum to energy of a single photon, $1/c$. Since a watt is 1 kg m^2 s^{-3}, and a newton is 1 kg m s^{-2}, then 1 second per meter = $1/3 \times 10^8$ newton per watt, which equals 3.3 micronewton per kilowatt (μN/kW). The limiting momentum per input power of a photon rocket is therefore 3.3 μN/kW. As many of these tests are at power levels in the tens or hundreds of watts, the propellantless devices appear to compare well with the photon rocket limit, yielding performance in the range of mN/kW.

As seen in the results in these Proceedings, there are some tantalizing indications of a verifiable signal in the Mach experiments. The devices tested and reported at Estes Park were older devices with known lower levels of thrust. The newer devices tested by Woodward typically have steady thrust signals of 2 μN and much higher transient thrust signals. The thrust signals are 3 to 5 sigma out of the noise in single runs, and more than 10 sigma when signal averaging is used to suppress the noise.

The EM-drive results remain somewhat more controversial among experimentalists. There is some concern among the community that all experimental artifacts have not been shown to be exhaustively eliminated, as would be necessary to establish confidence in a violation of momentum conservation in a closed, electromagnetic system. Soon after the Estes Park meeting, Hathaway undertook an assessment of the potential for experimental artifacts explaining EM-drive results of the Eagleworks group at Johnson Space Flight Center. The assessment did indeed find that insufficient control tests, of the kind Hathaway discusses in these Proceedings, were undertaken to establish confidence in the signals reported. So further control work may be needed.

The workshop provided some opportunity to develop consensus on the theoretical underpinnings of the Mach effect proposed by Woodward. There is agreement that conventional general relativity would seem to predict an effect of the sort upon which Woodward predicates his design. Tajmar and Williams both present alternative derivations in this Proceedings of a Mach effect from standard, linearized general relativity in the harmonic gauge. But where Woodward considered the inertial back-reaction force from all mass in the universe, and explicitly introduced a mechanical acceleration to obtain the effect, the separate but similar calculations by Williams and Tajmar did not consider either acceleration or back-reaction. While we remain optimistic, work is still to be done to couch a Mach effect within the framework of textbook general relativity.

Regarding the theoretical underpinnings of the EM-drive, there is consensus among experimentalists and theorists alike that no thrust should be possible from electromagnetic effects alone. This is reinforced by associated calculations with the COMSOL multi-physics numerical simulation tool reported by March, which showed zero net force from the microwaves.

Estes Park Advanced Propulsion Workshop

YMCA of the Rockies, Estes Park Center

Estes Park, Colorado

20-22 September 2016

Mission Statement

Our overarching objective is to find a way for humankind to reach the stars. This will require a breakthrough in propulsion.

The barrier to the stars can be conceived as two-fold: a fuel problem and a time-distance problem. We anticipate that both of these problems will entail a reckoning with general relativity. (see *Guidelines for Sessions* below for further detail).

Any breakthrough must be ultimately explainable to other technical people, must depart sensibly from the known laws of physics, and must involve verifiable experiments to test new effects. A discovery is not real until a second person understands it.

Conference Motivation

Fundamental research in physics funded by the NSF tends to focus on quantum gravity and string theory. NASA funded a small breakthrough propulsion program in the 1990s, but it was not sustained. Even if money was available, it is not clear how and where funds should be invested.

The typical conference format is ill-suited to this venture. Many conferences that accept breakthrough propulsion papers allow any person to pay a fee and present any dubious technical claim with little peer review or engagement with subject matter experts.

Therefore, we want to attempt to assemble a handful of potentially viable concepts for a propulsion breakthrough, and give each of them a rigorous, real-time, peer-review on the twin bases of theory and experiment. If someone has something with potential, they should relish a chance to explain it to others. If their scheme is ultimately not viable, they can be freed to join work in a more promising area.

Conference Style and Setting

Since conventional conference formats have not advanced a breakthrough in propulsion, we wanted to attempt something different: a "Shelter Island" approach. In 1947, Oppenheimer, Bethe, Feynman, and a dozen other luminaries of physics came together in an isolated retreat to tackle the major conceptual problems of quantum electrodynamics.

We feel such an environment can act as a reboot to the breakthrough propulsion effort, and re-establish it on a firm technical footing, grounded in known physical law and in scientific best practices. We hope our findings can be the nucleus of later efforts and can guide investment. Getting back on such a firm footing is best achieved at this time in an invitation-only event. We don't seek to limit participation, but merely to insure it is productive.

We have chosen to have the conference in historic Estes Park, Colorado, the week of 19 September, at the YMCA of the Rockies, Estes Park Center. It is a beautiful location in a rustic setting and should afford an opportunity to see the Colorado autumn colors in nearby Rocky Mountain National Park. A rate of $150/day includes all conference and meeting areas, lodging, and 3 meals per day. There is no conference fee.

Session Technical Format

The 3-day conference will address up to 12 concepts for a breakthrough in propulsion. We devote 2 hours per concept. The two hours are broken into theory and experiment sessions for the concept. The concept will be investigated on both grounds, with substantial give-and-take between the audience and the concept presenter, verbally and on the whiteboard.

The theory session will allow us to understand how the concept departs from existing theory of general relativity or electrodynamics. However, we are willing to entertain concepts with compelling experimental demonstrations for which there is yet no solid theory.

The experiment session will allow us to understand how to experimentally verify the concept, and the mechanism by which it could solve either the fuel or time-distance problems. However, we are willing to entertain concepts with compelling theoretical aspects well-rooted in known physics, for which concrete experiments could be contemplated.

We will moderate a disciplined and respectful interchange, working toward a goal of common understanding, while still "kicking all the tires" of rigorous peer-review.

Proceedings

The presentations and the discussions for the concept sessions will be recorded in a proceedings, along the lines of the Dirac birthday volume by Mehta, and other similar proceedings. We feel the technical discussion is an important part of the technical program to preserve. The intention is to create a conference proceedings that meets expected standards of peer-review. The proceedings will be made available to participants after the conference.

Guidelines for Concept Sessions

The concept should distinguish at the outset whether the work is in the area of the fuel problem, or of the time-distance problem. If that is an unwieldy binning, at least describe how your concept would get us to the stars.

The fuel problem is that conventional mass-ejection propulsion cannot power a craft to accelerate at 1 g for several years – a reasonable engineering goal – without impractical masses and volumes of fuel. We anticipate that this must amount to some form of breakthrough in gravity control, since the energy is spent to escape gravity. Even so, if the fuel problem were to be solved, the time-distance problem would still exist.

The time-distance problem is that civilization on earth can never colonize the galaxy if no signal can be sent faster than light, and if no object can be accelerated faster than light. The limiting speed of light, and the associated effects of relativistic time dilation, mean that we can never, for example, send scouting parties to the center of the galaxy and back. Our astronauts could go and come back – and they would see the center of the galaxy -- but time dilation would put them home in the distant future, long after their civilization died. So to find a solution to the time-distance problem is really to transcend the light barrier. To solve the time-distance problem would probably also imply a solution to the fuel problem.

A concept session should summarize its connection to the known laws of physics. For example, we would hope to see the extra terms in the Einstein field equations. If it is a wholly new theory, it should be covariant and follow from a Lagrangian.

A concept session should also summarize its experimental implications, or experimental set-up if a new effect has been detected. In either case, the goal is a verifiable experiment that can be reproduced by others.

We are also open to a promising theoretical framework without an obvious experiment to validate it – but the theory should have some potential for a testable prediction to falsify it. Likewise, we are open to an experimental effect without a theory to explain it – but the experiment should be rigorously described and verifiable.

We are looking forward to seeing you in September!

The Estes Park Workshop organizers
May 2016

www.ingramcontent.com/pod-product-compliance
Lightning Source LLC
Chambersburg PA
CBHW042055290426
44111CB00001B/10